The Behavior of Thin Walled Structures: Beams, Plates, and Shells

MECHANICS OF SURFACE STRUCTURES

Editors: W. A. Nash and G. Æ. Oravas

1. P. Seide, Small Elastic Deformations of Thin Shells. 1975
   ISBN 90-286-0064-7
2. V. Panc, Theories of Elastic Plates. 1975
   ISBN 90-286-0104-X
3. J. L. Nowinski, Theory of Thermoelasticity with Applications. 1978
   ISBN 90-286-0457-X
4. S. Lukasiewicz, Local Loads in Plates and Shells. 1979
   ISBN 90-286-0047-7
5. V. Fliřt, Statistics, Formfinding and Dynamics of Air-Supported Membrane
   Structures. 1983
   ISBN 90-247-2672-7
6. Yeh Kai-yuan, ed., Progress in Applied Mechanics. The Chien Wei-zang Anniversary
   Volume. 1986
   ISBN 90-247-3249-2
7. R. Negrutiu, Elastic Analysis of Slab Structures. 1986
   ISBN 90-247-3367-7
8. J. R. Vinson, The Behavior of Thin Walled Structures: Beams, Plates, and Shells.
   1989
   ISBN 90-247-3663-3

The Behavior of
Thin Walled Structures:
Beams, Plates, and Shells

By

Jack R. Vinson

Department of Mechanical Engineering
University of Delaware
Newark, Del., USA

Kluwer Academic Publishers

Dordrecht / Boston / London

Library of Congress Cataloging-in-Publication Data

```
Vinson, Jack R., 1929-
   The behavior of thin walled structures.

   (Mechanics of surface structures ; 8)
   Includes bibliographies and index.
   1. Thin-walled structures.  I. Title.  II. Series:
Mechanics of surface structures ; v. 8.
TA660.T5V56   1988         624.1'77        87-34773
```

ISBN 90-247-3663-3

Published by Kluwer Academic Publishers,
P.O. Box 17, 3300 AA Dordrecht, The Netherlands.

Kluwer Academic Publishers incorporates
the publishing programmes of
D. Reidel, Martinus Nijhoff, Dr W. Junk and MTP Press.

Sold and distributed in the U.S.A. and Canada
by Kluwer Academic Publishers,
101 Philip Drive, Norwell, MA 02061, U.S.A.

In all other countries, sold and distributed
by Kluwer Academic Publishers Group,
P.O. Box 322, 3300 AH Dordrecht, The Netherlands.

Printed in The Netherlands

To my beautiful wife, Midge, for her love,
encouragement, patience, and direct assistance
that made this textbook possible.

Contents

vii

Preface

This book is intended primarily as a teaching text, as well as a reference for individual study in the behavior of thin walled structural components. Such structures are widely used in the engineering profession for spacecraft, missiles, aircraft, land-based vehicles, ground structures, ocean craft, underwater vessels and structures, pressure vessels, piping, chemical processing equipment, modern housing, etc. It presupposes that the reader has already completed one basic course in the mechanics or strength of materials. It can be used for both undergraduate and graduate courses.

Since beams (columns, rods), plates and shells comprise components of so many of these modern structures, it is necessary for engineers to have a working knowledge of their behavior when these structures are subjected to static, dynamic (vibration and shock) and environmental loads.

Since this text is intended for both teaching and self-study, it stresses fundamental behavior and techniques of solution. It is not an encyclopedia of all research or design data, but provides the reader the wherewithal to read and study the voluminous literature.

Chapter 1 introduces the three-dimensional equations of linear elasticity, deriving them to the extent necessary to treat the following material. Chapter 2 presents, in a concise way, the basic assumptions and derives the governing equations for classical Bernoulli-Euler beams and plates in a manner that is clearly understood.

In Chapter 3, the solutions for beam problems are treated for a variety of commonly occurring static loads. In this chapter, Green's functions are developed and Galerkin's method are employed to illustrate these two powerful methods of solution.

In Chapter 4, both the Navier and Levy methods of solution for flat plates are shown, along with numerous solutions and tabulations of results for problems frequently encountered.

Because thermal loadings are so often involved in practical structures, a comprehensive discussion and treatment of thermal stresses and deformations is given in Chapter 5 for beams and plates, including the complexities of nonhomogeneous boundary conditions that result. The important fact that in many cases stresses due to thermal gradients are self equilibrating is illustrated. The analogous moisture effects associated with structures composed of polymeric materials is also discussed.

Chapter 6 introduces the theory and treats numerous problems associated with circular plates, because they are encountered in so many structural applications.

In Chapter 7 and 8, the eigenvalue problems of buckling and vibration are treated in a systematic way to enable the reader to analyze and determine critical buckling loads and natural frequencies to insure the structural integrity of thin walled columns, beams and plate structures.

Chapter 9 deals with energy methods for beams, columns, and plates in great detail. Energy solutions not only provide alternative approaches to solutions previously discussed, but provide the useful means to obtain approximate solutions to problems involving complicated loads or geometries which would be difficult or impossible to obtain otherwise.

Chapters 10, 11 and 12 introduce and treat many practical problems of cylindrical shells, which are one of the most commonly utilized shell configurations. Through mastering these chapters, the reader will understand the important effects of curvature and the 'bending boundary layer', and hence, can study the literature on other shell geometries more easily. Also, membrane theory, inextensional theory and Donnell's equations are treated as systematic approximations of the full set of equations.

In Appendix I, useful material property data are provided systematically for easy reference and use in solving problems for many metallic and polymeric isotropic materials at various temperatures.

Appendix 2 provides answers to many of the problems at the end of each chapter.

Throughout the text, the structures considered are of homogeneous isotropic materials, such as metals and polymers, in order that the concentration be on the structural mechanics, without the complications made necessary in dealing with structures composed of anisotropic composite materials. Those considerations are treated in other texts, such as *The Behavior of Structures Composed of Composite Materials* by Vinson and Sierakowski, published by Martinus Nijhoff, 1986.

Newark, Delaware   JACK R. VINSON

Equations of Linear Elasticity in Cartesian Coordinates

Sokolnikoff (Reference 1.2) derives in detail the formulation of the governing differential equations of elasticity in his first three chapters. This will not be repeated here, but rather the equations are presented and then utilized to systematically make certain assumptions in the process of deriving the governing equations for rectangular plates and beams.

1.1. Stresses

Consider an elastic body of any general shape. Consider the material to be a *continuum*, ignoring its crystalline structure and its grain boundaries. Also consider the continuum to be *homogeneous*, i.e., no variation of material properties with respect to the space coordinates. Then, consider a *material point* anywhere in the interior of the elastic body. If one assigns a Cartesian reference frame with axes x, y and z, shown in Figure 1.1, it is then convenient to assign a rectangular parallelepiped shape to the material point, and label it a *control element* of dimensions dx, dy and dz. The control element is defined to be infinitesimally small compared to the size of the elastic body, yet infinitely large compared to elements of the molecular structure, in order that the material can be considered a continuum.

On the surfaces of the control element there can exist both *normal stresses* (those perpendicular to the plane of the face) and *shear stresses* (those parallel to the plane of the face). On any one face these three stress components comprise a vector, called a *surface traction*.

It is important to note the sign convention and the subscript meaning of these surface stresses. For a stress component on a positive face, that is, a face whose outer normal is in the direction of a positive axis, that stress component is positive when it is directed in the direction of the positive axis. Conversely, when a stress is on a negative face of the control element, it is positive when it is directed in the negative axis direction. This procedure is followed in Figure 1.1. Also, the first subscript of any stress component on any face signifies the axis to which the outer normal is parallel. The second subscript refers to the axis to which that stress component is parallel. In the case of normal stresses the subscripts are seen to be repeated and often the two subscripts are shortened to one, i.e. $\sigma_i = \sigma_{ii}$ where $i = x$, y, or z.

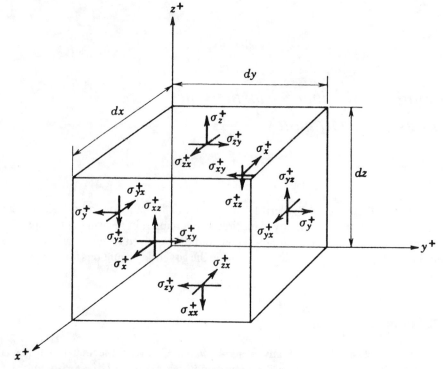

Figure 1.1. Control element in an elastic body showing positive direction of stresses.

1.2. Displacements

The displacements u, v and w are parallel to the x, y and z axes respectively and are positive when in the positive axis direction.

1.3. Strains

Strains in an elastic body are also of two types, extensional and shear. *Extensional strains*, ε_{ii}, where $i = x$, y, or z, are directed parallel to each of the axes respectively and are a measure of the change in dimension of the control volume in the subscripted direction due to the normal stresses acting on all surfaces of the control volume. Looking at Figure 1.2, one can define *shear strains*.

The shear strain γ_{ij} (where i and $j = x$, y, or z, and $i \neq j$) is a change of angle. As an example shown in Figure 1.2, in the x–y plane, defining γ_{xy} to be

$$\gamma_{xy} = \frac{\pi}{2} - \phi \quad \text{(in radians)} .\tag{1.1}$$

then,

$$\varepsilon_{xy} = \tfrac{1}{2}\gamma_{xy} .\tag{1.2}$$

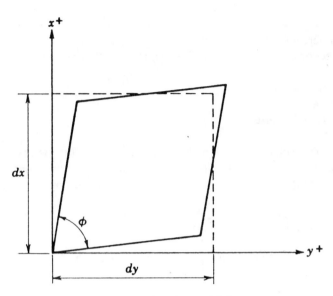

Figure 1.2. Shearing of a control element.

It is important to define the shear strain ε_{xy} to be one half the angle γ_{xy} in order to use tensor notation. However, in many texts and papers the shear strain is defined as γ_{xy}. Care must be taken to insure awareness of which definition is used when reading or utilizing a text or research paper, to obtain correct results in subsequent analysis. Sometimes ε_{ij} is termed tensor strain, and γ_{xy} is referred to as engineering strain (not a tensor quantity).

The rules regarding subscripts of strains are identical to those of stresses presented earlier.

1.4. Isotropy and Its Elastic Constants

An isotropic material is one in which the mechanical and physical properties do not vary with orientation. In mathematically modelling an isotropic material, the constant of proportionality between a normal stress and the resulting extension strain, in the sense of tensile tests is called the modulus of elasticity, E.

Similarly, from mechanics of materials, the proportionality between a shear stress and the resulting angle γ_{ij} described earlier, in a state of pure shear, is called the shear modulus, G.

One final quantity must be defined – the Poisson's ratio, denoted by v. It is defined as the ratio of the negative of the strain in the j direction to the strain in the i direction caused by a stress in the i direction, σ_{ii}. With this definition it is a positive quantity of magnitude $0 \leqslant v \leqslant 0.5$, for all isotropic materials.

The well known relationship between the modulus of elasticity, the shear modulus and Poisson's ratio should be remembered:

$$G = E/2(1 + v). \tag{1.3}$$

The basic equations of elasticity for a control element of an elastic body in a Cartesian reference frame can now be written. They are written in detail in the following sections, but the compact Einsteinian notation of tensor calculus is also provided.

1.5. Equilibrium Equations

A material point within an elastic body can be acted on by two types of forces: *body forces* (F_i) and surface tractions. The former are forces which are proportional to the mass, such as magnetic forces. The latter are stresses caused by neighboring control elements.

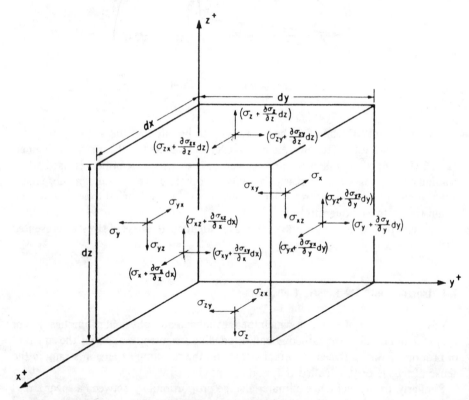

Figure 1.3. Control element showing variation of stresses.

Figure 1.1 is repeated above, but in Figure 1.3, the provision for stresses varying with respect to space is provided. Thus on the back face the stress σ_x is shown, while on the front face that stress value differs because σ_x is a function of x; hence, its value is $\sigma_x + (\partial \sigma_x / \partial x)\, dx$. Also shown are the appropriate expressions for the shear stresses.

The body forces per unit volume, $F_i (i = x, y, z)$ are proportional to mass and, because the body is homogeneous, are proportional to volume.

The summation of forces in the x direction can be written as

$$\left(\sigma_x + \frac{\partial \sigma_x}{\partial x} \, dx \right) dy \, dz + \left(\sigma_{yx} + \frac{\partial \sigma_{yx}}{\partial y} \, dy \right) dx \, dz$$

$$+ \left(\sigma_{zx} + \frac{\partial \sigma_{zx}}{\partial z} \, dz \right) dx \, dy - \sigma_x \, dy \, dz - \sigma_{yx} \, dx \, dz$$

$$- \sigma_{zx} \, dx \, dy + F_x \, dx \, dy \, dz = 0 . \tag{1.4}$$

After the cancellation, every term is multiplied by the volume, which upon division by the volume, results in

$$\frac{\partial \sigma_x}{\partial x} + \frac{\partial \sigma_{yx}}{\partial y} + \frac{\partial \sigma_{zx}}{\partial z} + F_x = 0 . \tag{1.5}$$

Likewise, in the y and z direction, the equilibrium equations are:

$$\frac{\partial \sigma_{xy}}{\partial x} + \frac{\partial \sigma_y}{\partial y} + \frac{\partial \sigma_{zy}}{\partial z} + F_y = 0 \tag{1.6}$$

$$\frac{\partial \sigma_{xz}}{\partial x} + \frac{\partial \sigma_{yz}}{\partial y} + \frac{\partial \sigma_z}{\partial z} + F_z = 0 . \tag{1.7}$$

In the compact Einsteinian notation, the above three equilibrium equations are written as

$$\sigma_{ki,k} + F_i = 0 \qquad (i, k = x, y, z) \tag{1.8}$$

where this is the ith equation, and the repeated subscripts k refer to each term be repeated in x, y and z, and where the comma means partial differentiation with respect to the subsequent subscript.

1.6. Stress-Strain Relations

The relationship between the stresses and strains at a material point in a three dimensional body mathematically describe the way the elastic material behaves. They are often referred to as the constitutive equations and are given below without derivation, because easy reference to many texts on strength of materials can be made.

$$\varepsilon_x = \frac{1}{E} \left[\sigma_x - v(\sigma_y + \sigma_z) \right] \tag{1.9}$$

$$\varepsilon_y = \frac{1}{E} \left[\sigma_y - v(\sigma_x + \sigma_z) \right] \tag{1.10}$$

$$\varepsilon_z = \frac{1}{E}\left[\sigma_z - v(\sigma_x + \sigma_y)\right] \tag{1.11}$$

$$\varepsilon_{xy} = \frac{1}{2G}\,\sigma_{xy} \tag{1.12}$$

$$\varepsilon_{yz} = \frac{1}{2G}\,\sigma_{yz} \tag{1.13}$$

$$\varepsilon_{zx} = \frac{1}{2G}\,\sigma_{zx} \tag{1.14}$$

From (1.9) the proportionality between the strain ε_x and the stress σ_x is clearly seen. It is also seen that stresses σ_y and σ_z affect the strain σ_x, due to the Poisson's ratio effect.

Similarly, in (1.12) the proportionality between the shear stress ε_{xy} and the shear stress σ_{xy} is clearly seen, the number 'two' being present due to the definition of ε_{xy} given in (1.2).

In the compact Einsteinian notation, the above six equations can be written as

$$\varepsilon_{ij} = a_{ijkl}\sigma_{kl} \tag{1.15}$$

where a_{ijkl} is the generalized compliance tensor.

1.7. Linear Strain-Displacement Relations

The strain-displacement relations are the kinematic equations relating the displacements that result from an elastic body being strained due to applied loads, or the strains that occur in the material when an elastic body is physically displaced.

$$\varepsilon_x = \frac{\partial u}{\partial x} \tag{1.16}$$

$$\varepsilon_y = \frac{\partial v}{\partial y} \tag{1.17}$$

$$\varepsilon_z = \frac{\partial w}{\partial z} \tag{1.18}$$

$$\varepsilon_{xy} = \frac{1}{2}\left(\frac{\partial u}{\partial y} + \frac{\partial v}{\partial x}\right) \tag{1.19}$$

$$\varepsilon_{xz} = \frac{1}{2}\left(\frac{\partial u}{\partial z} + \frac{\partial w}{\partial x}\right) \tag{1.20}$$

$$\varepsilon_{yz} = \frac{1}{2}\left(\frac{\partial v}{\partial z} + \frac{\partial w}{\partial y}\right) \tag{1.21}$$

In compact Einsteinian notation, these six equations are written as:

$$\varepsilon_{ij} = \tfrac{1}{2}(u_{i,j} + u_{j,i}) \qquad (i, j = x, y, z) \tag{1.22}$$

1.8. Compatibility Equations

The purpose of the compatibility equations is to insure that the displacements of an elastic body are single-values and continuous. They can be written as:

$$\frac{\partial^2 \varepsilon_{xx}}{\partial y\,\partial z} = \frac{\partial}{\partial x}\left(-\frac{\partial \varepsilon_{yz}}{\partial x} + \frac{\partial \varepsilon_{xx}}{\partial y} + \frac{\partial \varepsilon_{xy}}{\partial z} \right) \tag{1.23}$$

$$\frac{\partial^2 \varepsilon_{yy}}{\partial z\,\partial x} = \frac{\partial}{\partial y}\left(-\frac{\partial \varepsilon_{zx}}{\partial y} + \frac{\partial \varepsilon_{xy}}{\partial z} + \frac{\partial \varepsilon_{yz}}{\partial x} \right) \tag{1.24}$$

$$\frac{\partial^2 \varepsilon_{zz}}{\partial x\,\partial y} = \frac{\partial}{\partial z}\left(-\frac{\partial \varepsilon_{xy}}{\partial z} + \frac{\partial \varepsilon_{yz}}{\partial x} + \frac{\partial \varepsilon_{zx}}{\partial y} \right) \tag{1.25}$$

$$2\frac{\partial^2 \varepsilon_{xy}}{\partial x\,\partial y} = \frac{\partial^2 \varepsilon_{xx}}{\partial y^2} + \frac{\partial^2 \varepsilon_{yy}}{\partial x^2} \tag{1.26}$$

$$2\frac{\partial^2 \varepsilon_{yz}}{\partial y\,\partial z} = \frac{\partial^2 \varepsilon_{yy}}{\partial z^2} + \frac{\partial^2 \varepsilon_{zz}}{\partial y^2} \tag{1.27}$$

$$2\frac{\partial^2 \varepsilon_{zx}}{\partial z\,\partial x} = \frac{\partial^2 \varepsilon_{zz}}{\partial x^2} + \frac{\partial^2 \varepsilon_{xx}}{\partial z^2} \tag{1.28}$$

In compact Einsteinian notation, the compatibility equations are written as follows:

$$\varepsilon_{ij,kl} + \varepsilon_{kl,ij} - \varepsilon_{ik,jl} - \varepsilon_{jl,ik} = 0 \qquad (i, j, k, l = x, y, z). \tag{1.29}$$

However, in all of what follows, i.e., the analysis of beams, columns, plates, and shells, invariably the problems are placed in terms of displacements, and it is not necessary to utilize the compatibility equations.

1.9. Summary

It can be shown that both the stress and strain tensor quantities are symmetric, i.e.,

$$\sigma_{ij} = \sigma_{ji} \quad \text{and} \quad \varepsilon_{ij} = \varepsilon_{ji} \qquad (i, j = x, y, z). \tag{1.30}$$

Therefore, for the elastic solid there are fifteen independent variables; six stress components, six strain components and three displacements. In the case where compatibility is satisfied, there are fifteen equations: three equilibrium equations, six constitutive relations and six strain-displacement equations.

For a rather complete discussion of the equations of elasticity for anisotropic materials, see Reference 1.7, Chapter 2, Sections 2.1 through 2.3.

1.10. References

1.1. Timoshenko, S. and J. N. Goodier, *Theory of Elasticity*, New York: McGraw-Hill Book Company, 1970.
1.2. Sokolnikoff, I. S., *Mathematical Theory of Elasticity*, New York: McGraw-Hill Book Company, 2nd Edition, 1956.
1.3. Green, A. E. and W. Zerna, *Theoretical Elasticity*, Oxford University Press, 1954.
1.4. Love, A. E. H., *Mathematical Theory of Elasticity*, Cambridge University Press, 1934.
1.5. Muskhelishvili, N. I., *Some Basic Problems in the Mathematical Theory of Elasticity*, Noordhoff Publishing Company, 1953.
1.6. Love, A. E. H., *A Treatise on the Mathematical Theory of Elasticity*, New York: Dover Publications, Fourth Edition, 1944.
1.7. Vinson, J. R. and R. L. Sierakowski, *The Behavior of Structures Composed of Composite Materials*, Dordrecht: Martinus Nijhoff Publishers, 1986.

1.11. Problems

1.1. Prove that the stresses are symmetric, i.e., $\sigma_{ij} = \sigma_{ji}$.
   (Suggestion: take moments about the x, y, and z axes.)
1.2. When $v = 0.5$ a material is called 'incompressible'. Prove that for $v = 0.5$, under any set of stresses, the control volume of Figure 1.1 will not change volume.
1.3. An elastic body has the following strain field:

$$\varepsilon_{xx} = 2x^2 + 3xy + 4y^2 \qquad \varepsilon_{xy} = 0$$
$$\varepsilon_{yy} = x^2 - 2y^2 + z^2 \qquad \varepsilon_{yz} = 2y^2 - 3z^2$$
$$\varepsilon_{zz} = 2y^2 - z^2 \qquad \varepsilon_{xz} = 3z^2 - 2y^2$$

Does this strain field satisfy compatibility? Note: compatibility is not satisfied if any one or more of the compatibility equations is violated.

Derivation of the Governing Equations
for Beams and Rectangular Plates

The approach in this chapter is to systematically derive the governing equations for an isotropic classical, thin elastic rectangular plate, and subsequently simplify them for the governing equations of an isotropic elastic thin beam.

2.1. Assumptions of Plate Theory

In classical, linear thin plate theory, there are a number of assumptions that are necessary in order to reduce the three dimensional equations of elasticity to a two dimensional set that can be solved. Consider an elastic body shown in Figure 2.1, comprising the region $0 \leqslant x \leqslant a$, $0 \leqslant y \leqslant b$, and $-h/2 \leqslant z \leqslant h/2$, such that $h \ll a$ and $h \ll b$. This is called a plate.

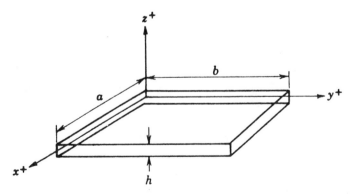

Figure 2.1. Rectangular plate.

The following assumptions are made.

1. A lineal element of the plate extending through the plate thickness, normal to the mid surface, x–y plane, in the unstressed state, upon the application of load:

9

    a. undergoes at most a translation and a rotation with respect to the original coordinate system;

    b. remains normal to the deformed middle surface.

2. A plate resists lateral and in-plane loads by bending, transverse shear stresses, and in-plane action, not through block like compression or tension in the plate in the thickness direction. This assumption results from the fact that $h/a \ll 1$ and $h/b \ll 1$.

From 1a the following is implied:

3. A lineal element through the thickness does not elongate or contract.
4. The lineal element remains straight upon load application.

In addition,

5. St. Venant's Principle applies.

    It is seen from 1a that the most general form for the two in-plane displacements is:

$$u(x, y, z) = u_0(x, y) + z\alpha(x, y) \tag{2.1}$$

$$v(x, y, z) = v_0(x, y) + z\beta(x, y) \tag{2.2}$$

where u_0 and v_0 are the in-plane middle surface displacements ($z = 0$), and α and β are rotations as yet undefined. Assumption 3 requires that $\varepsilon_z = 0$, which in turn means that the lateral deflection w is at most (from Equation 1.18)

$$w = w(x, y). \tag{2.3}$$

Also, Equation (1.11) is ignored.

    Assumption 4 requires that for any z, ε_{xz} = constant and ε_{yz} = constant at any specific location (x, y) on the plate middle surface for all z. Assumption 1b requires that the constant is zero, hence

$$\varepsilon_{xz} = \varepsilon_{yz} = 0 .$$

Assumption 2 means that $\sigma_z = 0$ in the stress strain relations.

    Incidentally, the assumptions above are identical to those of thin classical beam, ring and shell theory.

2.2. Derivation of the Equilibrium Equations for a Plate

Figure 2.2 shows the positive directions of stress quantities to be defined when the plate is subjected to lateral and in-plane loads.

The stress couples are defined as follows:

$$M_x = \int_{-h/2}^{+h/2} \sigma_x z \, dx \tag{2.4}$$

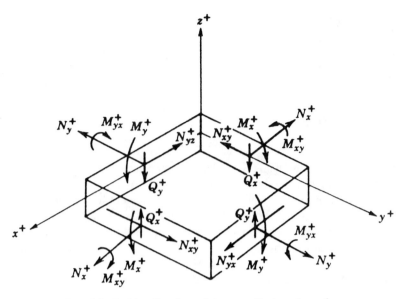

Figure 2.2. Positive directions of stress resultants and couples.

$$M_y = \int_{-h/2}^{+h/2} \sigma_y z \, dz \tag{2.5}$$

$$M_{xy} = \int_{-h/2}^{+h/2} \sigma_{xy} z \, dz \tag{2.6}$$

$$M_{yx} = \int_{-h/2}^{+h/2} \sigma_{yx} z \, dz = M_{xy} . \tag{2.7}$$

Physically, it is seen that the stress couple is the summation of the moment about the middle surface of all the stresses shown acting on all the infinitesimal control elements through the plate thickness at a location (x, y). In the limit the summation is replaced by the integration.

Similarly, the shear resultants are defined as,

$$Q_x = \int_{-h/2}^{+h/2} \sigma_{xz} \, dz \tag{2.8}$$

$$Q_y = \int_{-h/2}^{+h/2} \sigma_{yz} \, dz . \tag{2.9}$$

Again the shear resultant is physically the summation of all the shear stressed in the thickness direction acting on all of the infinitesimal control elements across the thick of the plate at the location (x, y).

Finally, the stress resultants are defined to be:

$$N_x = \int_{-h/2}^{+h/2} \sigma_x \, dz \tag{2.10}$$

$$N_y = \int_{-h/2}^{+h/2} \sigma_y \, dz \tag{2.11}$$

$$N_{xy} = \int_{-h/2}^{+h/2} \sigma_{xy} \, dz \tag{2.12}$$

$$N_{yx} = \int_{-h/2}^{+h/2} \sigma_{yx} \, dz = N_{xy} \tag{2.13}$$

These then are the sum of all the in-plane stresses acting on all of the infinitesimal control elements across the thickness of the plate at x, y.

Thus, in plate theory, the details of each control element under consideration are disregarded when one integrates the stress quantities across the thickness h. Instead of considering stresses at each material point one really deals with the integrated stress quantities defined above. The procedure to obtain the governing equations for plates from the equations of elasticity is to perform certain integrations on them.

Proceeding, multiply Equation (1.5) by $z \, dz$ and integrate between $-h/2$ and $+h/2$, as follows:

$$\int_{-h/2}^{+h/2} \left(z \, \frac{\partial \sigma_x}{\partial x} + z \, \frac{\partial \sigma_{xy}}{\partial y} + z \, \frac{\partial \sigma_{xy}}{\partial z} \right) dz = 0$$

$$\frac{\partial}{\partial x} \int_{-h/2}^{+h/2} \sigma_x z \, dz + \frac{\partial}{\partial y} \int_{-h/2}^{+h/2} \sigma_{xy} z \, dz + \int_{-h/2}^{+h/2} z \, \frac{\partial \sigma_{xz}}{\partial z} \, dz = 0$$

$$\left[\frac{\partial M_x}{\partial x} + \frac{\partial M_{xy}}{\partial y} + z \sigma_{xz} \right]_{-h/2}^{+h/2} - \int_{-h/2}^{+h/2} \sigma_{xz} \, dz = 0 \, .$$

In the above, the order of differentiation and integration can be reversed because x and z are orthogonal one to the other. Looking at the third term, $\sigma_{xz} = \sigma_{zx} = 0$ when there are no shear loads on the surface. This is not true for laminated plates. There, defining $\tau_{1x} = \sigma_{xz}(+h/2)$ and $\tau_{2x} = \sigma_{xz}(-h/2)$, the results are shown below in Equation (2.14). It should also be noted that for plates supported on an edge, σ_{xz} does not go to zero at $\pm h/2$, and so the theory is not accurate at that edge, but due to St. Venant's Principle, the solutions are satisfactory away from the edge supports.

$$\frac{\partial M_x}{\partial x} + \frac{\partial M_{xy}}{\partial y} + \frac{h}{2} \, (\tau_{1x} + \tau_{2x}) - Q_x = 0 \, . \tag{2.14}$$

Likewise Equation (1.6) becomes

$$\frac{\partial M_{xy}}{\partial x} + \frac{\partial M_y}{\partial y} + \frac{h}{2} \, (\tau_{1y} + \tau_{2y}) - Q_y = 0 \tag{2.15}$$

where

$$\tau_{1y} = \sigma_{xz}(+h/2) \quad \text{and} \quad \tau_{2y} = \sigma_{yz}(-h/2) \, .$$

These two equations describe the moment equilibrium of a plate element. Looking now at Equation (1.7), multiplying it by dz, and integrating between $-h/2$ and $+h/2$, results in

$$\int_{-h/2}^{+h/2} \left(\frac{\partial \sigma_{zx}}{\partial x} + \frac{\partial \sigma_{zy}}{\partial y} + \frac{\partial \sigma_z}{\partial z} \right) dz = 0$$

$$\frac{\partial Q_x}{\partial x} + \frac{\partial Q_y}{\partial y} + \sigma_z \int_{-h/2}^{+h/2} = 0$$

$$\frac{\partial Q_x}{\partial x} + \frac{\partial Q_y}{\partial y} + p_1(x, y) - p_2(x, y) = 0 \qquad (2.16)$$

where $p_1(x, y) = \sigma_z(+h/2)$, $p_2(x, y) = \sigma_z(-h/2)$.

One could also derive (2.16) by considering vertical equilibrium of a plate element shown in Figure 2.3.

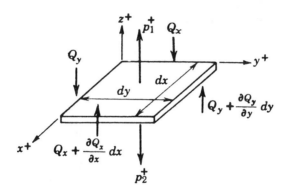

Figure 2.3. Vertical forces on a plate element.

One may ask why use is made of σ_z in this equation and not in the stress-strain relation? The foregoing is not realy inconsistent, since σ_z does not appear explicitly in Equation (2.16) and once away from the surface the normal surface traction is absorbed by shear and in-plane stresses rather than by σ_z in the plate interior, as stated previously in Assumption 2.

Similarly, multiplying Equation (1.5) and (1.6) by dz and integrating across the plate thickness results in the plate equilibrium equations in the x and y directions respectively, in terms of the in-plane stress resultants and the surface shear stresses.

$$\frac{\partial N_x}{\partial x} + \frac{\partial N_{xy}}{\partial y} + (\tau_{1x} - \tau_{2x}) = 0 \qquad (2.17)$$

$$\frac{\partial N_{xy}}{\partial x} + \frac{\partial N_y}{\partial y} + (\tau_{1y} - \tau_{2y}) = 0. \qquad (2.18)$$

2.3. Derivation of Plate Moment-Curvature Relations and Integrated Stress Resultant-Displacement Relations

Now, the plate equations must be derived corresponding to the elasticity stress strain relations. The strains ε_x, ε_y, and ε_{xy} will not be used explicitly since the stresses have been averaged by integrating through the thickness. Hence, displacements are utilized. Thus, combining (1.9) through (1.21) gives the following, remembering that σ_z has been assumed zero in the interior of the plate and excluding Equation (1.11) for reasons given previously:

$$\frac{\partial u}{\partial x} = \frac{1}{E} \left[\sigma_x - v\sigma_y \right] \tag{2.19}$$

$$\frac{\partial v}{\partial y} = \frac{1}{E} \left[\sigma_y - v\sigma_x \right] \tag{2.20}$$

$$\frac{1}{2} \left(\frac{\partial u}{\partial y} + \frac{\partial v}{\partial x} \right) = \frac{1}{2G} \sigma_{xy} \tag{2.21}$$

$$\frac{1}{2} \left(\frac{\partial v}{\partial z} + \frac{\partial w}{\partial y} \right) = \frac{1}{2G} \sigma_{yz} \tag{2.22}$$

$$\frac{1}{2} \left(\frac{\partial w}{\partial x} + \frac{\partial u}{\partial z} \right) = \frac{1}{2G} \sigma_{xz} . \tag{2.23}$$

Next, recall the form of the admissible displacements resulting from the plate theory assumptions:

$$u = u_0(x, y) + z\bar{\alpha}(x, y) \tag{2.24}$$

$$v = v_0(x, y) + z\beta(x, y) \tag{2.25}$$

$$w = w(x, y) \quad \text{only} . \tag{2.26}$$

In plate theory it is remembered that a lineal element through the plate will experience translations, rotations, but no extensions or contractions. For these assumptions to be valid, the lateral deflections are restricted to being small compared to the plate thickness. It is noted that if a plate is very thin, lateral loads can cause lateral deflections many times the thickness and the plate then behaves largely as a membrane because it has little or no bending resistance.

The assumptions of classical plate theory require that transverse shear deformation be zero. If $\varepsilon_{xz} = \varepsilon_{yz} = 0$ then from Equations (1.20) and (1.21).

$$\frac{1}{2} \left(\frac{\partial u}{\partial z} + \frac{\partial w}{\partial x} \right) = 0 \quad \text{or} \quad \frac{\partial u}{\partial z} = -\frac{\partial w}{\partial x} , \quad \text{likewise}$$

$$\frac{\partial v}{\partial z} = -\frac{\partial w}{\partial y} .$$

Hence, from Equations (2.25), (2.26) and the above, it is seen that the rotations are

$$\bar{\alpha} = -\frac{\partial w}{\partial x} \tag{2.27}$$

$$\beta = -\frac{\partial w}{\partial y} \tag{2.28}$$

Using (2.24) and (2.19), multiplying (2.19) through by $z\,dz$ and integrating from $-h/2$ to $+h/2$, one obtains

$$\int_{-h/2}^{h/2} \frac{\partial u_0}{\partial x} z\,dz + \int_{-h/2}^{+h/2} z^2 \frac{\partial \bar{\alpha}}{\partial x}\,dz = \int_{-h/2}^{+h/2} \frac{1}{E} [\sigma_x - v\sigma_y)z\,dz \,. \tag{2.29}$$

Likewise (2.25) and (2.20) result in

$$\int_{-h/2}^{h/2} \frac{\partial v_0}{\partial y} z\,dz + \int_{-h/2}^{+h/2} z^2 \frac{\partial \beta}{\partial y}\,dz = \int_{-h/2}^{+h/2} \frac{1}{E} [\sigma_y - v\sigma_x)z\,dz \tag{2.30}$$

and Equations (2.25), (2.25), and (2.21) give

$$\int_{-h/2}^{h/2} \left(\frac{\partial u_0}{\partial y} + \frac{\partial v_0}{\partial x} \right) z\,dz + \int_{-h/2}^{+h/2} \left(z^2 \frac{\partial \bar{\alpha}}{\partial y} + z^2 \frac{\partial \beta}{\partial x} \right) dz = \int_{-h/2}^{+h/2} \frac{1}{G} \sigma_{xy} z\,dz \,. \tag{2.31}$$

Integrating (2.29), (2.30), and (2.31) gives, using (2.27) and (2.28)

$$\frac{h^3}{12} \frac{\partial \bar{\alpha}}{\partial x} = \frac{1}{E} [M_x - vM_y] = -\frac{h^3}{12} \frac{\partial^2 w}{\partial x^2} \tag{2.32}$$

$$\frac{h^3}{12} \frac{\partial \beta}{\partial y} = \frac{1}{E} [M_y - vM_x] = -\frac{h^3}{12} \frac{\partial^2 w}{\partial y^2} \tag{2.33}$$

$$\frac{h^3}{12} \left(\frac{\partial \bar{\alpha}}{\partial y} + \frac{\partial \beta}{\partial x} \right) = \frac{1}{G} M_{xy} = -\frac{h^3}{6} \frac{\partial^2 w}{\partial x\, \partial y} \,. \tag{2.34}$$

Since $G = E/2(1 + v)$ $\tag{2.35}$

$$M_{xy} = -(1 - v)D \frac{\partial^2 w}{\partial x\, \partial y} \quad \text{where} \quad D = \frac{Eh^3}{12(1 - v^2)} \,. \tag{2.36}$$

Solving (2.33) and (2.34) for M_x and M_y results in,

$$M_x = -D \left[\frac{\partial^2 w}{\partial x^2} + v \frac{\partial^2 w}{\partial y^2} \right] \tag{2.37}$$

$$M_y = -D \left[\frac{\partial^2 w}{\partial y^2} + v \frac{\partial^2 w}{\partial x^2} \right] \,. \tag{2.38}$$

Equations (2.36) through (2.38) are known as the moment-curvature relations, and D is seen to be the flexural stiffness of the plate.

Likewise, substituting (2.37) and (2.38) into Equations (2.14) and (2.15) results in

$$Q_x = -D \frac{\partial}{\partial x} (\nabla^2 w) + \frac{h}{2} (\tau_{1x} + \tau_{2x}) \tag{2.39}$$

$$Q_y = -D \frac{\partial}{\partial y} (\nabla^2 w) + \frac{h}{2} (\tau_{1y} + \tau_{2y}) . \tag{2.40}$$

Also using Equations (2.24) and (2.25) substituting them into Equations (2.19) through (2.20), then multiplying the latter three equations by dz, integrating across the thickness, results in the following integrated stress-strain relationships:

$$N_x = K \left[\frac{\partial u_0}{\partial x} + v \frac{\partial v_0}{\partial y} \right] \tag{2.41}$$

$$N_y = K \left[\frac{\partial v_0}{\partial y} + v \frac{\partial u_0}{\partial x} \right] \tag{2.42}$$

$$N_{xy} = N_{yx} = Gh \left[\frac{\partial u_0}{\partial y} + \frac{\partial v_0}{\partial x} \right] , \tag{2.43}$$

where $Eh/(1 - v^2) = K$, the plate extensional stiffness. Equations (2.41) through (2.43) describe the in-plane force and deformation behavior.

2.4. Derivation of the Governing Equations for a Plate

The equations governing the lateral deflections, and the bending and shearing action of a plate can be summarized as follows:

$$\frac{\partial M_x}{\partial x} + \frac{\partial M_{xy}}{\partial y} - Q_x + \frac{h}{2} (\tau_{1x} + \tau_{2x}) = 0 \tag{2.44}$$

$$\frac{\partial M_{xy}}{\partial x} + \frac{\partial M_y}{\partial y} - Q_y + \frac{h}{2} (\tau_{1y} + \tau_{2y}) = 0 \tag{2.45}$$

$$\frac{\partial Q_x}{\partial x} + \frac{\partial Q_y}{\partial y} + p_1 - p_2 = 0 \tag{2.46}$$

$$M_x = -D \left[\frac{\partial^2 w}{\partial x^2} + v \frac{\partial^2 w}{\partial y^2} \right] \tag{2.47}$$

$$M_y = -D \left[\frac{\partial^2 w}{\partial y^2} + v \frac{\partial^2 w}{\partial x^2} \right] \tag{2.48}$$

$$M_{xy} = -D(1 - v) \frac{\partial^2 w}{\partial x \, \partial y} . \tag{2.49}$$

The equations governing the in-plane stress resultants and in-plane midsurface displacements are:

$$\frac{\partial N_x}{\partial x} + \frac{\partial N_{xy}}{\partial y} + (\tau_{1x} - \tau_{2x}) = 0 \tag{2.50}$$

$$\frac{\partial N_{xy}}{\partial x} + \frac{\partial N_y}{\partial y} + (\tau_{1y} - \tau_{2y}) = 0 \tag{2.51}$$

$$N_x = K\left[\frac{\partial u_0}{\partial x} + v\,\frac{\partial v_0}{\partial y}\right] \tag{2.52}$$

$$N_y = K\left[\frac{\partial v_0}{\partial y} + v\,\frac{\partial u_0}{\partial x}\right] \tag{2.53}$$

$$N_{xy} = Gh\left[\frac{\partial u_0}{\partial y} + \frac{\partial v_0}{\partial x}\right]. \tag{2.54}$$

It should be noted that in classical, thin plate theory the equations related to bending and shear, Equations (2.44) through (2.49), are completely uncoupled from the equations dealing with in-plane loads and displacements, equations (2.50) through (2.54). [*Note*: in Chapter 7, we shall see that when inplane loads are highly compressive, the in-plane loads do indeed cause lateral displacements (buckling), but a more sophisticated theory will be evolved at that time.]

It should also be noted that the flexural stiffness D of the plate corresponds closely to the EI in beam theory, but is in terms of a unit width, and incorporates the Poisson's ratio effect. Likewise a similar correspondence exists between the extensional stiffness K and the EA in beam theory.

Equations (2.44) through (2.54) are the eleven governing plate equations. First note that the plate can only tell the difference between normal tractions on the upper and lower surface. Hence, one can define $p(x, y)$ as

$$p_1(x, y) - p_2(x, y) = p(x, y). \tag{2.55}$$

Substituting (2.44) and (2.45) in (2.46) results in the following for the case of no shear stresses on the plate upper and lower surfaces:

$$\frac{\partial^2 M_x}{\partial x^2} + 2\,\frac{\partial^2 M_{xy}}{\partial x\,\partial y} + \frac{\partial^2 M_y}{\partial y^2} + p(x, y) = 0. \tag{2.56}$$

Substituting (2.47) through (2.49) in this results in:

$$D\left[\frac{\partial^4 w}{\partial x^4} + 2\,\frac{\partial^4 w}{\partial x^2\,\partial y^2} + \frac{\partial^4 w}{\partial y^4}\right] = p(x, y)$$

$$D\nabla^4 w = p(x, y), \tag{2.57}$$

where

$$\nabla^2(\ ) = \frac{\partial^2(\ )}{\partial x^2} + \frac{\partial^2(\ )}{\partial x^2} \quad \text{and} \quad \nabla^4(\ ) = \nabla^4(\nabla^2(\ )).$$ (2.58)

∇^2 is really the sum of the curvatures in two orthogonal directions at the location x, y in the plate. ∇^4 is physically, then, the sum of the curvatures of the sum of the curvatures in orthogonal directions. One might say that it is a measure of 'bulginess'.

Next, treating Equations (2.50) through (2.54) by substituting Equations (2.52) through (2.54) into the two equilibrium equations, becomes, after considerable manipulation, and for the case of no surface shear stresses,

$$\nabla^4 u_0 = 0$$ (2.59)

$$\nabla^4 v_0 = 0.$$ (2.60)

For the bending vibrations of a plate, an inertial load per unit platform area is added as an equivalent force per unit area, resulting in Equation (2.57) being modified, as seen below:

$$D\nabla^4 w = p - \rho h \frac{\partial^2 w}{\partial t^2}$$ (2.61)

where ρ is the *mass* density of the plate material, and t is the coordinate of time. Here, $w = w(w, y, t)$ and $p = p(x, y, t)$. This modification can be made because the theory is linear and superposition is possible.

In a plate of varying thickness, $h = h(x, y)$, the following equation is derived rather than (2.57):

$$\nabla^2(D\nabla^2 w) - (1 - v)\Diamond^4(D, w) = p(x, y)$$ (2.62)

where $\Diamond^4$ is the die operator defined as

$$\Diamond^4(D, w) = \frac{\partial^2 D}{\partial x^2} \frac{\partial^2 w}{\partial y^2} - 2 \frac{\partial^2 D}{\partial x \, \partial y} \frac{\partial^2 w}{\partial x \, \partial y} + \frac{\partial^2 D}{\partial y^2} \frac{\partial^2 w}{\partial x^2}.$$ (2.63)

If a plate is on an elastic foundation, in which a linear foundation modulus k, in units of lbs/in/in^2 can be defined, then Equation (2.57) is altered by adding in the additional lateral force per unit platform area:

$$D\nabla^4 w = p(x, y) - kw.$$ (2.64)

Since classical linear elasticity is involved herein, superposition permits the writing of a vibrating plate on an elastic foundation.

$$D\nabla^4 w = p(x, y, t) - \rho h \frac{\partial^2 w}{\partial t^2} - kw.$$ (2.65)

2.5. Boundary Conditions

First, the boundary conditions for the bending of a plate subjected to lateral loads, Equation (2.57) will be discussed. Additional boundary conditions for Equations (2.59) and (2.60) for a plate subjected also to in-plane loads or displacements will be discussed later.

Since (2.57) is a fourth order partial differential equation in x and y describing the bending of a plate, four boundary conditions are needed on the x edges and four are needed on the y edges, i.e., two on each edge. For the clamped and simply supported edges, knowledge of beam theory dictates the following:

For a clamped edge   *For a simply supported edge*

$$w = 0 \qquad\qquad\qquad w = 0$$

(2.66)

$$\frac{\partial w}{\partial n} = 0 \qquad\qquad\qquad M_n = 0$$

where n is the direction normal to the edge.

For a Free Edge

Consider an x = constant free edge. Since by definition a free edge has no loads applied to it, Figure 2.2 shows that M_x, M_{xy}, and Q_x all are zero on that edge. Hence, six boundary conditions must be satisfied on the two x = constant plate edges. However, the plate equation is only fourth order in x, hence one cannot specify more than two boundary conditions on each edge. [*Note*: In a more advanced plate theory that includes the effects of transverse shear deformation, $\varepsilon_{xz} \neq 0$ and $\varepsilon_{yz} \neq 0$, the governing equations are sixth order in both x and y and the problem discussed here does not occur.]

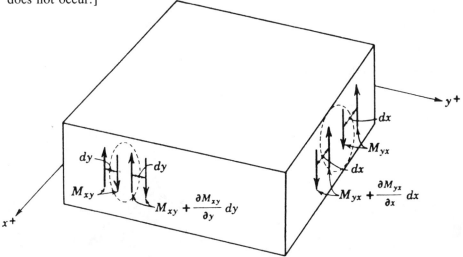

Figure 2.4. Development of the Kirchhoff boundary conditions for a free edge.

To eliminate the problem, Kirchhoff proceeded as follows: On the free $x =$ constant edge, M_x is set equal to zero. M_{xy}, the twisting stress couple is considered to be a couple consisting of two forces of magnitude M_{xy} separated by a small distance dy, as shown in Fig. 2.4. Since the stress couple M_{xy} is not constant in general along an edge, nearby is another couple, $M_{xy} + (\partial M_{xy}/\partial y)\,dy$. It too can be regarded as two forces of magnitude $M_{xy} + (\partial M_{xy}/\partial y)\,dy$, separated by a distance dy. Therefore, considering an infinitesimal region of the edge shown within the dotted line, it is seen that there is a force M_{xy} positive downward, a force $M_{xy} + (\partial M_{xy}/\partial y)\,dy$ positive upwards as well as the force due to the transverse shear resultant, $Q_x\,dy$, acting positive upwards. These must equal zero, hence,

$$- M_{xy} + M_{xy} + \frac{\partial M_{xy}}{\partial y}\,dy + Q_x\,dy = 0$$

or

$$V_x = Q_x + \frac{\partial M_{xy}}{\partial y} = 0 \tag{2.67}$$

where V_x is called the effective shear resultant on a free edge.

Physically it is seen that on the free edge Q_x nor M_{xy} are zero, only the relationship given by (2.67) is zero. However, this approximation was found to have sufficient accuracy that it has been widely used in plate analysis and is known as the Kirchhoff Boundary Condition.

Likewise on a $y =$ constant free edge

$$V_y = Q_y + \frac{\partial M_{xy}}{\partial x} = 0 \tag{2.68}$$

and of course on either edge the other boundary condition is

$$M_n = 0 \tag{2.69}$$

where n refers to the directional normal to the edge.

Edge Elastically Supported Against Deflection

Suppose there exists a linear spring support at an edge of magnitude c lbs/in^2. Then the boundary conditions become:

$$M_n = 0 \tag{2.70}$$

$$V_n + cw = 0 \tag{2.71}$$

or

$$Q_n + \frac{\partial M_{ns}}{\partial s} + cw = 0$$

or

$$\frac{\partial^3 w}{\partial n^3} + (2 - v) \frac{\partial^3 w}{\partial n \, \partial s^2} - \frac{cw}{D} = 0 \tag{2.72}$$

where s refers to the direction parallel to the edge.

Edge Elastically Restrained Against Rotation

Suppose there exists a torsional spring support at an edge of r in lbs/in. Then the boundary conditions would be:

$$V_n = 0 \tag{2.73}$$

$$M_n + r \frac{\partial w}{\partial n} = 0$$

or

$$\frac{\partial^2 w}{\partial n^2} + v \frac{\partial^2 w}{\partial s^2} - \frac{r}{D} \frac{\partial w}{\partial n} = 0 . \tag{2.74}$$

In-Plane Boundary Conditions

In Section 2.4, it was seen that the governing equations involving the in-plane forces and midsurface displacements are completely uncoupled from the other equations, the boundary conditions for which have been discussed above.

In the case that a plate is not subjected to any prescribed in-plane loads or prescribed midsurface displacements, the solutions to Equations (2.59) and (2.60) are

$$u_0 = v_0 = 0 .$$

For other cases, the details of the edge conditions of the plate structure being analyzed must be studied in detail, to specify which boundary conditions should be prescribed. However through the use of variational procedures, which will be discussed in Chapter 9, it can be shown that the boundary conditions to use in solving Equations (2.59) and (2.60) are:

For an x = constant edge:

Either u_0 is prescribed or $N_x = 0$
and
Either v_0 is prescribed or $N_{xy} = 0$

For a y = constant edge:

Either v_0 is prescribed or $N_y = 0$
and
Either u_0 is prescribed or $N_{yx} = 0$.

2.6. Stress Distribution within a Plate

In plate theory because all equations are integrated across the thickness only integrated stress quantities are obtained. For stresses on a control element or material point within a plate, one must *assume* a stress distribution. This is done by means of an analogy to beam theory. Thus,

$$\sigma_x = \frac{M_x z}{h^3/12} + \frac{N_x}{h} \tag{2.75}$$

$$\sigma_y = \frac{M_y z}{h^3/12} + \frac{N_y}{h} \tag{2.76}$$

$$\sigma_{xy} = \frac{M_{xy} z}{h^3/12} + \frac{N_{xy}}{h} \tag{2.77}$$

$$\sigma_{xz} = \frac{3Q_x}{2h}\left[1 - \left(\frac{z}{h/2}\right)^2\right] - \frac{S_x}{4} \tag{2.78}$$

$$\sigma_{yz} = \frac{3Q_y}{2h}\left[1 - \left(\frac{z}{h/2}\right)^2\right] - \frac{S_y}{4} \tag{2.79}$$

where

$$S_x = \tau_{1x}\left[1 - 2\left(\frac{z}{h/2}\right) - 3\left(\frac{z}{h/2}\right)^2\right]$$

$$+ \tau_{2x}\left[1 + 2\left(\frac{z}{h/2}\right) - 3\left(\frac{z}{h/2}\right)^2\right] \tag{2.80}$$

$$S_y = \tau_{1y}\left[1 - 2\left(\frac{z}{h/2}\right) - 3\left(\frac{z}{h/2}\right)^2\right]$$

$$+ \tau_{2y}\left[1 + 2\left(\frac{z}{h/2}\right) - 3\left(\frac{z}{h/2}\right)^2\right] \tag{2.81}$$

It can easily be shown that these distributions satisfy the definitions of Equations (2.4) through (2.13). Equally important they satisfy the equilibrium equations of elasticity (1.5) and (1.6) exactly, and Equation (1.7) on the average. Thus the stresses obtained through the use of plate theory (or beam, shell and ring theory) are not exact, in the sense of being three dimensional elasticity theory solutions, but they are very close to the exact solution.

2.7. References

2.1.   Sokolnikoff, I. S., *Mathematical Theory of Elasticity*, New York: McGraw-Hill Book Company, Inc., 2nd Edition, 1956.

2.2. Timoshenko, S. and A. Woinowsky-Krieger, *Theory of Plates and Shells*, New York: McGraw-Hill Book Company, Inc., 2nd Edition, 1959.

2.3. Marguerre, K. and H. T. Woernle, *Elastic Plates*, Blaisdell Publishing Company, 1970.

2.4. Mansfield, E. H., *The Bending and Stretching of Plates*, Pergamon Press, 1964.

2.5. Jaeger, L. G., *Elementary Theory of Elastic Plates*, Pergamon Press, 1964.

2.6. Morley, L. S. D., *Skew Plates and Structures*, Pergamon Press, 1963.

2.7. Vlasov, V. Z. and N. N. Leont'ev, *Beams, Plates and Shells on Elastic Foundations*, published for NASA and NSF by the Israel Program for Scientific Translations, 1966.

2.8. Vinson, J. R., *Structural Mechanics: The Behavior of Plates and Shells*, New York: John Wiley and Sons, 1974.

2.9. Vinson, J. R. and T. W. Chou, *Composite Materials and Their Use in Structures*, London: Applied Science Publishers, 1975.

2.10. Vinson, J. R. and R. L. Sierakowski, *The Behavior of Structures Composed of Composite Materials*, Dordrecht: Martinus Nijhoff Publishers, 1986.

2.8. Problems

2.1. The governing equations for a rectangular plate subjected to a laterial distributed load $p(x, y)$ are given by Equations (2.57) through (2.60). However, when the plate is subjected to surface shear stresses τ_{1x}, τ_{2x}, τ_{1y}, and τ_{2y}, additional terms are added which are functions of those surface shear stresses, such that the equations can be written as:

$$D\nabla^4 w = p(x, y) + e(\tau_{1x}, \tau_{2x}, \tau_{1y}, \tau_{2y})$$

$$K\nabla^4 u_0 = f(\tau_{1x}, \tau_{2x}, \tau_{1y}, \tau_{2y})$$

$$K\nabla^4 v_0 = g(\tau_{1x}, \tau_{2x}, \tau_{1y}, \tau_{2y}).$$

Find the functions e, f, and g.

2.2. Derive equations (2.72), starting with Equations (2.54) through (2.59).

2.3. Show that the stress distributions of Sections (2.6) do in fact satisfy the definitions of Equations (2.4) through (2.13).

2.4. Show that the stress distributions of Section 2.6 satisfy Equations (1.5) and (1.6) where the body forces $F_i = 0$. Do they satisfy Equations (1.7) with $F_z = 0$? Do they satisfy (1.7) on the average, i.e.,

$$\frac{1}{h} \int_{-h/2}^{+h/2} \left[\frac{\partial \sigma_{xz}}{\partial x} + \frac{\partial \sigma_{yz}}{\partial y} + \frac{\partial \sigma_z}{\partial z} \right] dz = 0 \ ?$$

2.5. Starting with the pertinent elasticity equations, derive Equations (2.50) and (2.52).

2.6. Consider the plate shown in Figure 2.1. The plate is subjected to a constant in-plane load in the y-direction, $N_y = N_0$, only.

   a. What are the stresses σ_x, σ_y and σ_{xy} in the plate?

   b. What are the displacements u, v and w in the plate? Assume $v_0 = 0$ along the $y = 0$ edge, and $u_0 = 0$ along the $x = 0$ edge.

Beams and Rods

3.1. General Remarks

This chapter provides the means by which to analyze beams, rods, and columns. These structural members have a commonality in that in a Cartesian coordinate system, one dimension, the length (L), is an order of magnitude greater than the other two dimensions, the width (b) and the height (h). Hence for a beam, rod, or column:

$$\frac{h}{L} \ll 1 \quad \text{and} \quad \frac{b}{L} \ll 1 . \tag{3.1}$$

A beam is a structure that is primarily subjected to a lateral load resulting in bending of the beam. A rod is such a structure subjected to a load in the length direction resulting in tensile stresses, while a column is this type of structure subjected to a compressive load in the length direction, such that the induced stresses are compressive. However, in this chapter, the column studied is one in which the loads only cause compressive stresses; when the loads increase, there will occur a critical load at which time the compressive loads in the length direction will cause lateral displacements, a phenomenon called buckling or elastic instability. This phenomenon is treated in detail in Chapter 7, because it involves a theory more complicated than that involved in this chapter.

3.2. Development of the Governing Equations

Consider the beam-rod-column shown in Figure 3.1 below.

This chapter follows the chapter on three-dimensional elasticity and the chapter on plates, because beam-rod-column theory is a logical simplification of the two-dimensional equations for a plate.

It is assumed herein that for a beam, the lateral loads, distributed or concentrated, are in the x–z plane; while the tensile or compressive loads for the rod or column are in the x direction. Therefore all action, loads, and responses are in the x–z plane, and no quantity varies in the y-direction, and there are no stresses, strains, dis-

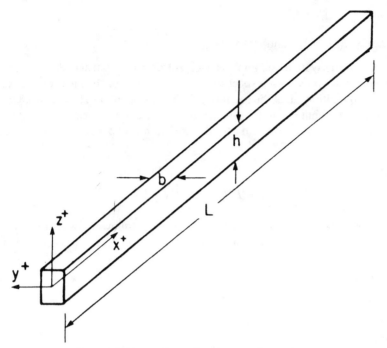

Figure 3.1. Nomenclature for a beam, rod, or column.

placements, or loads in the y-direction, i.e.,

$$N_y = N_{xy} = M_y = M_{xy} = Q_y = \tau_{1y} = \tau_{2y} = v_0 = 0 . \tag{3.2}$$

From Chapter 2, the equilibrium equations for a plate reduce to:

$$\frac{dN_x}{dx} = \tau_{1x} - \tau_{2x} = 0 \tag{3.3}$$

$$\frac{dQ_x}{dx} + p_1(x) - p_2(x) = \frac{dQ_x}{dx} + p(x) = 0 \tag{3.4}$$

$$\frac{dM_x}{dx} - Q_x + \frac{h}{2}[\tau_{1x} + \tau_{2x}] = 0 . \tag{3.5}$$

Likewise, the integrated stress strain relations would be reduced to:

$$N_x = K \frac{du_0}{dx} \tag{3.6}$$

$$M_x = -D \frac{d^2 w}{dx^2} \tag{3.7}$$

$$Q_x = -D \frac{d^3 w}{dx^3} \tag{3.8}$$

where

$$K = Eh/(1 - v^2) \quad \text{and} \quad D = Eh^3/12(1 - v^2).$$ (3.9)

Notice that the derivatives are total derivatives rather than partial derivatives because all functions vary only with respect to x. Also because of this, it is convenient and traditional to multiply all of the equations in (3.3) through (3.9) by the width b. Along with this is the definition of an in-plane load resultant P, a beam bending moment M_b, a beam shear resultant V, and a transverse beam load per unit length $q(x)$, defined as:

$$P = N_x b, \quad M_b = M_x b, \quad V = Q_x b, \quad q(x) = p(b)b$$

$$\text{and} \quad Kb = EA, \quad Db = EI \quad \text{where} \quad v = 0.$$ (3.10)

Substituting (3.10) into (3.3) through (3.9) gives the following, where for simplicity only, the surface shear stress terms τ_{1x} and τ_{2x} are omitted (but easily retained).

$$\frac{dP}{dx} = 0$$ (3.11)

$$\frac{dV}{dx} + q(x) = 0$$ (3.12)

$$\frac{dM_b}{dx} - V = 0$$ (3.13)

$$P = EA \, \frac{du_0}{dx}$$ (3.14)

$$M_b = -EI \, \frac{d^2 w}{dx^2}$$ (3.15)

$$V = -EI \, \frac{d^3 w}{dx^3}.$$ (3.16)

Note, that as in the case of plates, the equations governing in-plane loads (3.11) and (3.14) are completely uncoupled from the equations involving lateral loads, lateral displacements, bending moments, and shear resultants. [*Note*: this is not the case when buckling is considered, using a more refined theory, in Chapter 7.]

From (3.11), it is seen that P = constant, determined by the applied axial load at the ends, and therefore the substitution of (3.14) into (3.11), and integrating the result gives:

$$u_0(x) = \left(\frac{P}{EA} \right) x + C_0$$ (3.17)

where C_0 is a constant of integration determined by the location where $u_0(x)$ is specified.

The substitution of (3.16) into (3.12) provides the governing differential equation for the bending of a beam:

$$EI \, \frac{d^4 w}{dx^4} = q(x) . \tag{3.18}$$

Note that in (3.17), A is the cross-sectional area of the rod or column, and in (3.18), I is the first area of inertia of the cross-section with respect to the neutral surface of bending. For a rectangular section of Figure 3.1, $A = bh$ and $I = bh^3/12$. Those quantities for other sections are catalogued in many references.

3.3. Solutions for the Beam Equation

From (3.18), progressive integration provides the following general solutions for a lateral distributed load, $q(x)$:

$$\frac{d^4 w}{dx^4} = \frac{q(x)}{EI} \tag{3.19}$$

$$\frac{d^3 w}{dx^3} = \frac{1}{EI} \int q(x) \, dx + C_1 \tag{3.20}$$

$$\frac{d^2 w}{dx^2} = \frac{1}{EI} \int \int q(x) \, dx \, dx + C_1 x + C_2 \tag{3.21}$$

$$\frac{dw}{dx} = \frac{1}{EI} \int \int \int q(x) \, dx \, dx \, dx + \frac{C_1 x^2}{2} + C_2 x + C_3 \tag{3.22}$$

$$w(x) = \frac{1}{EI} \int \int \int \int q(x) \, dx \, dx \, dx \, dx + \frac{C_1 x^3}{6} + \frac{C_2 x^2}{2} + C_3 x + C_4 . \tag{3.23}$$

In the above, there are four constants of integration (C_1, C_2, C_3, and C_4) which must be solved for a set of boundary conditions prescribed. For the classical boundary conditions:

Simple-supported Edge

$$w = M_b = 0 \tag{3.24}$$

Clamped Edge

$$w = \frac{dw}{dx} = 0 \tag{3.25}$$

Free Edge

$$M_b = V = 0 \quad \text{or} \quad \frac{d^2 w}{dx^2} = \frac{d^3 w}{dx^3} = 0 . \tag{3.26}$$

Other boundary conditions are discussed in Chapter 2.

3.4. Stresses in Beams – Rods – Columns

The stresses in these structural members are easily obtained from those of Chapter 2 for plates, utilizing the definitions of Section (3.2):

$$\sigma_x = \frac{P}{A} + \frac{M_b z}{I} \tag{3.27}$$

$$\sigma_{xz} = \frac{3V}{2A}\left[1 - \left(\frac{z}{h/2}\right)^2\right]. \tag{3.28}$$

These would have additional terms, analogs to those of Chapter 2 if imposed surface shear stresses τ_{1x} and τ_{2x} exist.

3.5. Example: Clamped-Clamped Beam with a Constant Lateral Load, $q(x) = -q_0$

This is illustrated in Figure 3.2.

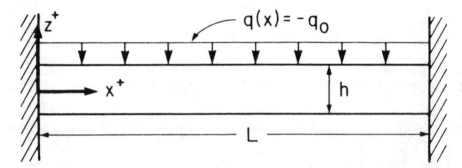

Figure 3.2. Clamped-clamped beam with a uniform lateral load.

From (3.16) and (3.17)

$$P = 0, \quad C_0 = 0, \quad u_0(x) = 0. \tag{3.29}$$

From (3.19) through (3.23).

$$C_3 = C_4 = 0 \tag{3.30}$$

$$C_1 = \frac{q_0 L}{2EI}, \quad C_2 = -\frac{q_0 L^2}{12EI}$$

$$w(x) = -\frac{q_0}{24EI}(x^4 - 2Lx^3 + L^2 x^2). \tag{3.31}$$

Here, as is always true, the lateral deflection $w(x)$ is proportional to the applied load, and inversely proportional to the flexural stiffness, EI.

From (3.31), the maximum deflection can be found; the result is:

$$w_{\max} = w(L/2) = -\frac{q_0 L^4}{384 EI}. \tag{3.32}$$

Using the derivatives of (3.31), or using (3.21) with (3.30), the location and magnitude of the maximum stress is determined. Here it is found that

$$\sigma_{x_{\max}} = \sigma_x \begin{pmatrix} 0 \\ L \end{pmatrix}, \pm h/2 = \pm \frac{q_0 L^2}{2Ah}. \tag{3.33}$$

Likewise, the location and magnitude of the maximum shear stress can be found easily.

3.6. Example: Cantilevered Beam with a Uniform Lateral Load $q(x) = -q_0$

This is illustrated in Figure 3.3.

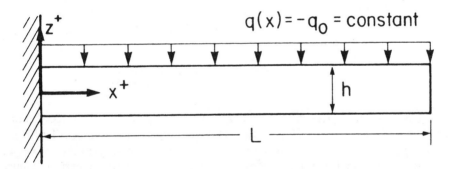

Figure 3.3. Cantilevered beam subjected to a uniform lateral load.

From (3.16) and (3.17),

$$P = C_0 = u_0(x) = 0. \tag{3.34}$$

From (3.19) through (3.23),

$$C_1 = \frac{q_0 L}{EI}, \quad C_2 = -\frac{q_0 L^2}{2EI}$$

$$C_3 = C_4 = 0 \tag{3.35}$$

$$w(x) = -\frac{q_0}{24EI}(x^4 - 4x^3 L + 6L^2 x^2). \tag{3.36}$$

Again, the maximum deflection and stresses are:

$$w_{max} = w(L) = -\frac{q_0 L^4}{8EI}$$ (3.37)

$$\sigma_{x_{max}} = \sigma_x(0, \pm h/2) = \pm \frac{3q_0 L^2}{Ah} .$$ (3.38)

3.7. Example: Simply Supported Beam with a Uniform Load over Part of Its Length

Consider the following example (Figure 3.4):

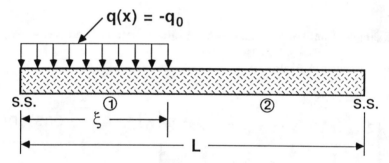

Figure 3.4. Uniform beam with discontinuous load.

One complication arises when the beam of flexural stiffness EI over the entire length is subjected to a distributed load $q(x)$ over only one portion of its length. In this example, $q(x) = -q_0$ over $0 \leqslant x \leqslant \xi$, and $q(x) = 0$ over the portion $\xi \leqslant x \leqslant L$. Because of this load discontinuity, the beam must be divided into two parts as shown in Figure 3.4. In what follows the beam equations under the load will be subscripted 1, and those dealing with the unloaded portion are subscripted 2. The boundary conditions at the ends can be easily written as follows:

$$w_1(0) = 0 \qquad\qquad w_2(L) = 0$$ (3.39)

$$M_{b_1}(0) = \frac{d^2 w_1(0)}{dx^2} = 0 \qquad M_{b_2} = \frac{d^2 w_2(L)}{dx^2} = 0 .$$

Four other boundary conditions must be determined, and they describe the compatibilities that must exist at the junction of beam parts 1 and 2, namely

$$w_1(\xi) = w_2(\xi)$$ (3.40)

$$\frac{dw_1(\xi)}{dx} = \frac{dw_2(\xi)}{dx}$$ (3.41)

$$M_{b_1}(\xi) = M_{b_2}(\xi) \quad \text{or} \quad \frac{d^2 w_1(\xi)}{dx^2} = \frac{d^2 w_2(\xi)}{dx^2} \tag{3.42}$$

$$V_1(\xi) = V_2(\xi) \quad \text{or} \quad \frac{d^3 w_1(\xi)}{dx^3} = \frac{d^3 w_2(\xi)}{dx^3} . \tag{3.43}$$

Using the above, it is seen that there are eight equations to determine the eight boundary constants C_{11}, C_{21}, C_{31}, C_{41}, C_{12}, C_{22}, C_{32}, and C_{42} in Equations (3.20) through (3.23), where the second subscript is used to denote Section 1 or Section 2. The equations to solve involve solving an 8×8 set of algebraic equations and lengthy manipulations. The final constants in this example are:

$$C_{11} = \frac{q_0}{EI} \left[\xi - \frac{\xi^2}{2L} \right] \qquad C_{12} = -\frac{q_0 \xi^2}{2EIL}$$

$$C_{21} = 0 \qquad C_{22} = \frac{q_0 \xi^2}{2EI}$$

$$C_{31} = \frac{q_0}{24EI} \left(-4L\xi^2 - \frac{\xi^4}{L} + \xi^3 \right) \qquad C_{32} = -\frac{q_0}{24EI} \left[4L\xi^2 + \frac{\xi^4}{L} \right]$$

$$C_{41} = 0 \qquad C_{42} = \frac{q_0 \xi^4}{24EI} .$$

The resulting displacement equations are:

$$w_1(x) = -\frac{q_0}{24EI} \left[x^4 - 4x^3\xi + \frac{4x^3\xi^2}{L} + 4Lx\xi^2 + \frac{x\xi^4}{L} - x\xi^3 \right], \quad 0 \leqslant x \leqslant \xi \tag{3.44}$$

$$w_2(x) = -\frac{q_0}{24EI} \left[\frac{2x^3\xi^2}{L} - 6x^2\xi^2 + 4Lx\xi^2 - \frac{x\xi^4}{L} - \xi^4 \right], \quad \xi \leqslant x \leqslant L \tag{3.45}$$

The location and magnitude of the maximum deflection will occur in either Section 1 or 2 depending upon the extent of the load, where ξ is located. Then by taking second derivatives of the deflections function one can determine the bending moments in Sections 1 and 2, and subsequently determine the locations and magnitude of the maximum stress for a given ξ.

From (3.44) and (3.45), it can be seen that if $\xi = 0$, i.e. no load, the beam consists only of Section 2, and since each term has ξ in the numerator, $w = 0$ everywhere. Similarly, if $\xi = L$, the beam only consists of Section 1, the problem is that of a simple supported beam subjected to a uniform load over the entire length of $q(x) = -q_0$, with the result that:

$$w(x) = -\frac{q_0}{24EI} \left[x^4 - 2Lx^3 + L^3 x \right] \tag{3.46}$$

wherein

$$w_{max} = w(L/2) = -\frac{5q_0L^4}{384EI}.$$ (3.47)

Subsequently,

$$\sigma_x(x) = \frac{M_b(x)z}{I} = \tfrac{1}{2}(x^2 - Lx)\frac{q_0z}{I}$$ (3.48)

and

$$\sigma_{x_{max}} = \sigma_x(L/2, \pm h/2) = \mp\frac{q_0L^2}{Ah}.$$ (3.49)

3.8. Beam with an Abrupt Change in Stiffness

Consider a beam shown in Figure 3.5, wherein the dimensions change from h_1 to h_2, hence in the Section 1, the flexural stiffness is EI_1, on the right EI_2.

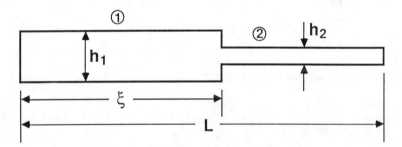

Figure 3.5. Beam of discontinuous flexural stiffness.

Again, Equations (3.20) through (3.23) are used for each section, satisfying C_{11} through C_{42}, as in the previous section by solving for the boundary conditions at each end, and the four matching boundary conditions at $x = \xi$, wherein for Section 1, EI_1 is used and for the right section EI_2 is used.

It is now seen that in general, for each load discontinuity and/or each structural discontinuity, the beam must be divided into sections involving continuous loads and continuous geometry, with rational matching of conditions at each boundary. For example, if there were three load discontinuities and two abrupt changes in geometry, all at different axial locations, the beam must be divided into six sections, and twenty-four equations must be solved to obtain twenty-four boundary value constants. Then, an investigation must be made of candidate sections to determine in which the maximum deflections and in which the maximum stress occurs. This can become quite lengthy.

Another approach to these complicated problems will be illustrated in Chapter 9, Energy Methods.

3.9. Beam Subjected to Concentrated Loads

Consider a beam of length L, subjected to a concentrated load P located at $x = \xi$, as shown in Figure 3.6.

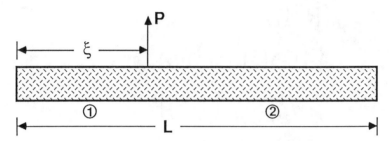

Figure 3.6. Beam subjected to a concentrated lateral load.

Again, because of the discontinuity, the beam must be divided into sections at each discontinuity, such as Section 1 from $0 \leqslant x \leqslant \xi$, and Section 2 at $\xi \leqslant x \leqslant L$, as shown.

Here, there is no lateral distributed load, $q(x)$, so only the homogeneous solution of (3.23) is used for each of the two sections, or

$$w_i(x) = \frac{C_{1i}x^3}{6} + \frac{C_{2i}x^2}{2} + C_{3i}x + C_{4i} \quad (i = 1, 2). \tag{3.50}$$

For this class of problems, again there are eight boundary conditions; four to be determined to satisfy the boundary conditions at the ends of the beam, and four to properly match conditions at the load P.

At the ends, the boundary conditions such as simple support, clamped, free, etc. can be imposed using equations (3.24) through (3.26) for example.

At $x = \xi$, the load position, it is clear that for Sections 1 and 2, the deflection, slope, and bending moment must match. Therefore,

$$w_1(\xi) = w_2(\xi) \tag{3.51}$$

$$\frac{dw_1(\xi)}{dx} = \frac{dw_2(\xi)}{dx} \tag{3.52}$$

$$M_{b_1}(\xi) = M_{b_2}(\xi) \quad \text{or} \quad \frac{d^2w_1(\xi)}{dx^2} = \frac{d^2w_2(\xi)}{dx^2}. \tag{3.53}$$

Looking at the vertical equilibrium, it can be seen that the last boundary condition is

$$V_1(\xi) - V_2(\xi) = P \quad \text{or} \quad -\frac{d^3w_1(\xi)}{dx^3} + \frac{d^3w_2(\xi)}{dx^3} = \frac{P}{EI} \,. \tag{3.54}$$

Using (3.19) through (3.23) to satisfy the above boundary conditions, the solutions for a beam with various boundary conditions are:

Beam Clamped at Each End

$$w_1(x) = \frac{P\xi}{2EI}\left[1 - \left(\frac{\xi}{L}\right)^2\right]x^2$$

$$-\frac{P}{6EI}\left[1 - 3\left(\frac{\xi}{L}\right)^2 + 2\left(\frac{\xi}{L}\right)^3\right]x^3, \quad 0 \leqslant x \leqslant \xi \tag{3.55}$$

$$w_2(x) = -\frac{P\xi^3}{6EI} + \frac{P\xi^2 x}{2EI} + \frac{P\xi x^2}{2EI}\left[-2 + \frac{\xi}{L}\right]$$

$$+\frac{Px^3}{6EI}\left(\frac{\xi}{L}\right)^2\left[3 - 2\left(\frac{\xi}{L}\right)\right], \quad \xi \leqslant x \leqslant L\,. \tag{3.56}$$

Beam Clamped at $x = 0$, Simply Supported at $x = L$

$$w_1(x) = \frac{PL}{4EI}\left(\frac{\xi}{L}\right)\left[2 - 3\left(\frac{\xi}{L}\right) + \left(\frac{\xi}{L}\right)^2\right]x^2$$

$$-\frac{P}{12EI}\left[2 - 3\left(\frac{\xi}{L}\right)^2 + \left(\frac{\xi}{L}\right)^3\right]x^3, \quad 0 \leqslant x \leqslant \xi \tag{3.57}$$

$$w_2(x) = -\frac{P\xi^3}{6EI} + \frac{P\xi^2 x}{2EI} - \frac{PLx^2}{4EI}\left(\frac{\xi}{L}\right)^2\left[3 - \left(\frac{\xi}{L}\right)\right]$$

$$+\frac{Px^3}{12EI}\left(\frac{\xi}{L}\right)^2\left[3 - \frac{\xi}{L}\right], \quad \xi \leqslant x \leqslant L\,. \tag{3.58}$$

Beam Clampled at $x = 0$, Free at $x = L$

$$w_1(x) = \frac{P\xi x^2}{2EI} - \frac{Px^3}{2EI}, \quad 0 \leqslant x \leqslant \xi \tag{3.59}$$

$$w_2(x) = -\frac{P\xi x^3}{6EI} + \frac{P\xi^2 x}{2EI}, \quad \xi \leqslant x \leqslant L\,. \tag{3.60}$$

Beam Simply Supported at Each End

$$w_1(x) = \frac{PL^2 x}{6EI}\left(\frac{\xi}{L}\right)\left[2 + \left(\frac{\xi}{L}\right) - 3\left(\frac{\xi}{L}\right)^2\right] - \frac{Px^3}{6EI}\left[1 - \frac{\xi}{L}\right], \quad 0 \leqslant x \leqslant \xi \tag{3.61}$$

$$w_2(x) = -\frac{P\xi^3 x}{6EI} + \frac{PL^2 x}{6EI}\left(\frac{\xi}{L}\right)\left[2 + \frac{\xi}{L}\right] - \frac{P\xi x^2}{2EI} + \frac{Px^3}{6EI}\left(\frac{\xi}{L}\right), \quad \xi \leqslant x \leqslant L. \quad (3.62)$$

These equations are quite general for a beam of uniform flexural stiffness EI, subjected to any load P acting at $x = \xi$. For instance (3.59) can be used for a clamped-free beam at load P at the tip $x = L$ of by letting $\xi = L$.

Also because the equations evolve from linear theory, superposition can be used if there are two concentrated loads at two different locations, but care must be used to accurately depict regions to the left and right of each load to insure correct solutions.

After the solutions $w_i(x)$ are found, equations (3.15) and (3.16) are used to obtain bending moments and shear resultants, and (3.27) and (3.28) used to determine stresses everywhere.

3.10. Solutions by Green's Functions

From Section 3.9, consider a beam subjected to a unit concentrated load, i.e., $P = 1$. By way of example, consider the beam simply supported at each end given by (3.61) and (3.62). In this case,

$$w_1(x) = \frac{L^2 x}{6EI}\left(\frac{\xi}{L}\right)\left[2 + \left(\frac{\xi}{L}\right) - 3\left(\frac{\xi}{L}\right)^2\right] - \frac{x^3}{6EI}\left[1 - \left(\frac{\xi}{L}\right)\right] =$$

$$= G_1(x, \xi), \quad 0 \leqslant x \leqslant \xi \qquad (3.63)$$

$$w_2(x) = -\frac{\xi^3}{6EI} + \frac{L^2 x}{6EI}\left(\frac{\xi}{L}\right)\left[2 + \frac{\xi}{L}\right] - \frac{\xi x^2}{2EI} + \frac{x^3}{6EI}\left(\frac{\xi}{L}\right) =$$

$$= G_2(x, \xi), \quad \xi \leqslant x \leqslant L. \qquad (3.64)$$

Note that $G_1(x)$ is the deflection to the left of the load, $G_2(x)$ to the right of the load.

Green's functions G_1 and G_2 are the deflection at x due to a unit load at ξ.

It can be reasoned that any distributed load $q(x)$ is in fact an infinity of concentrated loads, which can be summed to obtain the solution of the response to a distributed load. Because of the infinity of concentrated loads, the infinite summation can be replaced by an integration such that the following is correct.

$$w(x) = \int_x^L G_1(x, \xi)q(\xi)\,d\xi + \int_0^x G_2(x, \xi)q(\xi)\,d\xi. \qquad (3.65)$$

For a beam simply supported at each end (3.63) and (3.64) would be inserted for G_1 and G_2, and other analogous expressions from Section (3.9) would be used for beams with other boundary conditions.

Solving (3.65), (3.63), and (3.64) with $q(x) = -q_0$, a uniform load over the entire length results in

$$w(x) = -\frac{q_0}{24EI}(x^4 - 2Lx^3 + L^3 x) \qquad (3.66)$$

which is the solution obtained through solving the governing differential Equation (3.18) and satisfying boundary conditions. Likewise using the Green's function approach shown above for a uniform load results in (3.31) for the clamped-clamped beam, and (3.36) for the cantilevered beam.

Thus, with the use of Green's functions one uses an integral equation where the Green's functions satisfy the boundary conditions, rather than solving the governing differential equation and matching boundary conditions. This alternative approach

(1) is general for any governing differential equation such as a beam, plate, or shell, for example.
(2) can save great computational difficulty for complicated problems.

As an example of the latter, consider the following problem (Figure 3.7):

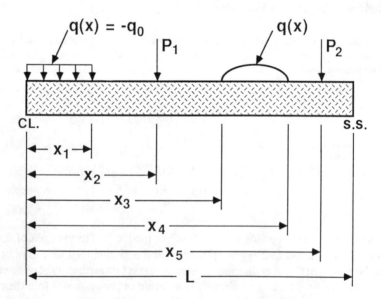

Figure 3.7. Beam subjected to a variety of lateral loads.

In the above, to solve the problem through solving the governing differential equations and matching boundary conditions involves dividing the beam into six sections, wherein twenty-four boundary value constants must be solved for, and particular solutions obtained for each of the distributed loads. As a result, $w(x)$ is found everywhere after all constants are solved for simultaneously.

Using the Green's function approach, suppose one hypothesizes, for example, that the maximum deflection and bending moment occurs in the region $x_3 \leqslant x \leqslant x_4$. One obtains $G_1(x, \xi)$ and $G_2(x, \xi)$ for the clamped-simply supported beam from (3.63) and (3.64). Then for $x_3 \leqslant x \leqslant x_4$, $w(x)$ is given by

$$w(x) = \int_0^{x_1} G_2(x, \xi)(-q_0)\,\mathrm{d}\xi - G_2(x, x_2)P_1$$

$$- \int_{x_3}^{x} G_2(x, \xi)q(\xi)\,\mathrm{d}\xi - \int_{x}^{x_4} G_1(x, \xi)q(\xi)\,\mathrm{d}\xi$$

$$- G_1(x, x_s)P_2 \quad \text{for} \quad x_3 \leqslant x \leqslant x_4. \tag{3.67}$$

To check that the maxima occur in this region, one could easily investigate adjoining regions, i.e., $x_2 \leqslant x \leqslant x_3$, and $x_4 \leqslant x \leqslant x_5$. Having obtained (3.67), maximum deflections and stresses are determined straightforwardly.

3.11. Tapered Beam Solution Using Galerkin's Method

In previous sections, methods to deal with concentrated loads, and discontinuities in either loads or cross-sections were discussed. However, when there is a continuous change in cross-section, the beam equation of (3.18) is seen to be

$$\frac{\mathrm{d}^2}{\mathrm{d}x^2}\left[EI(x)\,\frac{\mathrm{d}^2w}{\mathrm{d}x^2}\right] = q(x) \tag{3.68}$$

where here $I(x)$ is a continuously varying function of x. Depending on what that function of x is, the solution could be rather complicated and/or tedious.

For these kinds of problems, Galerkin's method is well suited. No attempt is made here to provide a detailed comprehensive introduction to Galerkin's method which is well treated in numerous other texts. However, consider an ordinary differential equation (although the method is equally useful for partial differential equations as well as nonlinear equations as shown below).

$$L[w(x)] - q(x) = 0 \tag{3.69}$$

where L is any differential operator, and $q(x)$ is a forcing function. Boundary conditions must be homogeneous; if not, a transformation of variables must be made to attain homogeneous boundary conditions.

In Galerkin's method one assumes a complete set of coordinate functions $X_n(x)$, $n = 1, 2, 3, \ldots, N$, which satisfies the prescribed homogeneous boundary conditions. Although these functions need not be an orthogonal set, if they are, the procedure is much simpler. Therefore, assume

$$w_n(x) = \sum_{n=1}^{N} C_n X_n(x) \tag{3.70}$$

where the C_n are constants.

By using only the first N terms, an error $\varepsilon_N(x)$ can be defined from (3.70) and (3.69), as

$$\varepsilon_N(x) = L(w_N) - q(x). \tag{3.71}$$

If $\varepsilon_N(x)$ is sufficiently small, then $W_n(x)$ is considered to be a satisfactory approximation to $w(x)$. So $\varepsilon_n(x)$ can be viewed as an error function, and the task is to select proper C_n to minimize $\varepsilon_n(x)$. In the Galerkin method, this is done by the following orthogonality condition:

$$\int_0^L \varepsilon_N(x) X_m(x)\, dx = 0 \quad \text{where} \quad 1 \leqslant m \leqslant N. \tag{3.72}$$

This is equivalent to minimizing the mean square error and insures that $W_n(x)$ will converge to $W(x)$ in the mean.

From above this is equivalent to stating:

$$\int_0^L \left[L \left(\sum_{n=1}^N C_n X_n(x) - q(x) \right) \right] X_m(x)\, dx = 0, \quad m = 1, 2, 3, \ldots, N. \tag{3.73}$$

Therefore, (3.73) is a set of N algebraic equations through which one can determine the C_n to minimize $\varepsilon_n(x)$. This is a powerful technique.

Now consider a tapered beam subjected to a lateral distributed load $q(x)$. Assume the variable stiffness can be written as:

$$EI(x) = EI_0 f(x). \tag{3.74}$$

Equation (3.68) can then be written as

$$(fw'')'' = \frac{q(x)}{EI_0} \tag{3.75}$$

where primes denote differentiation with respect to x. Hence,

$$fw^{iv} + 2f' w''' + f'' w'' = \frac{q(x)}{EI_0} \tag{3.76}$$

If the plate is simply supported, i.e., $w = w'' = 0$ at $x = 0, L$, coordinate functions can be chosen as follows:

$$X_n(x) = \sin \frac{n\pi x}{L} = \sin \alpha_n x \tag{3.77}$$

where $\alpha_n = n\pi/L$.

These are a complete set of functions which satisfy the boundary conditions. Thus,

$$w_n(x) = \sum_{n=1}^\infty C_n \sin \alpha_n x \tag{3.78}$$

and from (3.71)

$$\varepsilon_n(x) = \sum_{n=1}^\infty \{ C_n f \alpha_n^4 \sin \alpha_n x - 2C_n f' \alpha_n^3 \cos \alpha_n x$$

$$- f'' C_n \alpha_n^2 \sin \alpha_n x \} - \frac{q(x)}{EI_0}. \tag{3.79}$$

From (3.72) and (3.73), Galerkin's procedure requires

$$\sum_{n=1}^{N} C_n \left\{ \int_0^L f\alpha_n^4 \sin \alpha_n x \sin \alpha_m x \, dx \right.$$

$$\left. - \int_0^L f' \alpha_n^3 \cos \alpha_n x \sin \alpha_m x \, dx - \int_0^L f'' \alpha_n^2 \sin \alpha_n x \sin \alpha_m x \, dx \right\}$$

$$- \int_0^L \frac{q(x)}{EI_0} \sin \alpha_m x \, dx = 0 . \tag{3.80}$$

To continue the example, let $f(x) = 1 + (x/L)$, then (3.80) becomes:

$$\sum_{n=1}^{N} C_n \left\{ \int_0^L \alpha_n^4 \left(1 + \frac{x}{L}\right) \sin \alpha_n x \sin \alpha_m x \, dx \right.$$

$$\left. - \int_0^L \left(\frac{1}{L}\right) 2\alpha_n^3 \cos \alpha_n x \sin \alpha_m x \, dx \right\}$$

$$- \int_0^L \frac{q(x)}{EI_0} \sin \alpha_m x \, dx = 0 . \tag{3.81}$$

For $n \neq m$

$$\int_0^L \left(1 + \frac{x}{L}\right) \sin \alpha_n x \sin \alpha_m x \, dx = \frac{[(-1)^{n+m} - 1]}{(\alpha_n^2 - \alpha_m^2)^2} \frac{2\alpha_n \alpha_m}{L}$$

$$\int_0^L \cos \alpha_n x \sin \alpha_m x \, dx = \frac{\alpha_m}{(\alpha_n^2 - \alpha_m^2)} [(-1)^{n+m} - 1]$$

So (3.81) is, for $n \neq m$

$$\sum_{n=1}^{N} C_n \left\{ \frac{2\alpha_n^5 \alpha_m}{L} \frac{[(-1)^{n+m} - 1]}{(\alpha_n^2 - \alpha_m^2)^2} - \frac{2\alpha_n^3 \alpha_m}{L} \frac{[(-1)^{n+m} - 1]}{\alpha_n^2 - \alpha_m^2} \right\} =$$

$$\int_0^L \frac{q(x)}{EI_0} \sin \alpha_m x \, dx .$$

For $n = m$

$$\int_0^L \left(1 + \frac{x}{L}\right) \sin \alpha_n x \sin \alpha_m x \, dx = 3L/4$$

$$\int_0^L \cos \alpha_n x \sin \alpha_m x \, dx = 0 .$$

Thus, (3.81) becomes for $n = m$

$$\sum_{n=1}^{N} C_n \left(\frac{3L\alpha_n^4}{4}\right) = \int_0^L \frac{q(x)}{EI_0} \sin \alpha_m x \, dx .$$

If N is taken as 3, the final set of three non-homogeneous algebraic equations are written as follows to obtain C_1, C_2, and C_3. The first, second, and third equations are for $m = 1, 2, 3$, respectively.

$$C_1 \left(\frac{3L\alpha_1^4}{4} \right) + C_2 \left[\frac{2\alpha_2^5 \alpha_1 (-2)}{L(\alpha_2^2 - \alpha_1^2)^2} - \frac{2\alpha_2^3 \alpha_1 (-2)}{L(\alpha_2^2 - \alpha_1^2)} \right] = \int_0^L \frac{q(x)}{EI_0} \sin \alpha_1 x \, dx$$

$$C_1 \left[\frac{2\alpha_1^5 \alpha_2 (-2)}{L(\alpha_1^2 - \alpha_2^2)^2} - \frac{2\alpha_1^3 \alpha_2 (-2)}{L(\alpha_1^2 - \alpha_2^2)} \right] + C_2 \left(\frac{3L\alpha_2^4}{4} \right)$$

$$+ C_3 \left[\frac{2\alpha_3^5 \alpha_2 (-2)}{L(\alpha_3^2 - \alpha_2^2)^2} - \frac{2\alpha_3^3 \alpha_2 (-2)}{L(\alpha_3^2 - \alpha_2^2)} \right] = \int_0^L \frac{q(x)}{EI_0} \sin \alpha_2 x \, dx$$

$$C_2 \left[\frac{2\alpha_2^5 \alpha_3 (-2)}{L(\alpha_2^2 - \alpha_3^2)^2} - \frac{2\alpha_2^3 \alpha_3 (-2)}{L(\alpha_2^2 - \alpha_3^2)} \right] + C_3 \left(\frac{3L\alpha_3^4}{4} \right) = \int_0^L \frac{q(x)}{EI_0} \sin \alpha_3 x \, dx \, .$$

To complete the problem for any distributed load, $q(x)$, is straightforward.

To know what value of N to take can only be determined by solving the problem with one higher integer and determining that the previous approximation is sufficient.

One thing to remember is that the result of this method is the determination of an approximate deflection $w(x)$. To determine stresses requires solving for w'' since the bending stresses are proportional to w''. Taking derivatives of an approximative function causes an increase in the error through differentiation. Hence, for stress critical structural members, N must be determined to suitably approximate the maximum stress. This may require a higher value of N than that required to achieve a certain accuracy in maximum deflection.

3.12. Problems

3.1. Consider the cantilevered beam of Section 3.6. If the beam is $1''$ wide, $15''$ long, composed of steel where $E = 30 \times 10^6$ psi, $v = 0.3$, an allowable stress of 50 000 psi, and an applied constant lateral load of 30 lbs/in. of length, what thickness h is required to prevent overstressing or a tip deflection to not exceed $0.3''$?

3.2. Consider a beam of constant cross-sectional area A, clamped at $x = 0$ and simply supported at $x = L$. If the beam is subjected to a uniform lateral load $q(x) = -q_0$,
   a. Where and what is the maximum stress?
   b. Where and what is the maximum deflection?

Solutions to Problems of Rectangular Plates

4.1. Some General Solutions to the Biharmonic Equation

The governing equation for a rectangular plate subjected to lateral distributed loads is given by

$$\nabla^4 w = \frac{\partial^4 w}{\partial x^4} + \frac{2\partial^4 w}{\partial x^2\,\partial y^2} + \frac{\partial^4 w}{\partial y^4} = \frac{p(x,\,y)}{D}\,. \tag{4.1}$$

First, the homogeneous equation, $\nabla^4 w = 0$ is investigated. It is interesting to do this in order to identify the functions that are characteristic of the biharmonic equation.

One of the most common methods used to solve this homogeneous equation is by separation of variables. This process can be attempted when the boundary conditions are homogeneous. We cannot count upon the separation of variables to yield all of the complete exact solutions, but it will give all the separable solutions. There *may* be others.

Let

$$w(x,\,y) = X(x)\,Y(y)\,. \tag{4.2}$$

From (4.1) and (4.2),

$$X^{iv}Y + 2X''\,Y'' + XY^{iv} = 0\,.$$

Dividing by XY gives

$$\frac{X^{iv}}{X} + 2\,\frac{X''}{X}\frac{Y''}{Y} + \frac{Y^{iv}}{Y} = 0\,. \tag{4.3}$$

The variables are still not separated, hence, let

$$\frac{X^{iv}}{X} = f(x)\,, \quad \frac{X''}{X} = g(x)\,, \quad \frac{Y''}{Y} = k(y)\,, \quad \frac{Y^{iv}}{Y} = p(y)\,.$$

Equation (4.3) becomes

$$f(x) + 2g(x)k(y) + p(y) = 0\,. \tag{4.4}$$

Differentiating with respect to x gives,

$$f'(x) + 2g'(x)k(y) = 0$$

or,

$$\frac{f'(x)}{g'(x)} + 2k(y) = 0.$$

For this to be true, then $f'(x)/g'(x)$ = constant and $k(y)$ = constant. Thus,

$$- \lambda^2 = k(y) = \text{constant}. \tag{4.5}$$

Similarly differentiating (4.4) with respect to y gives

$$2g(x)k'(y) + p'(y) = 0$$

$$\frac{p'(y)}{k'(y)} + 2g(x) = 0.$$

Hence, the following must be true.

$$- \gamma^2 = g(x) = \text{constant}. \tag{4.6}$$

Case 1 $k(y) = - \lambda^2$

   Case 1a $\lambda^2 > 0$

$$k(y) = \frac{Y''}{Y} = - \lambda^2$$

$$Y'' + \lambda^2 Y = 0 \quad \text{or} \quad Y = \begin{Bmatrix} \cos \lambda y \\ \sin \lambda y \end{Bmatrix}.$$

Substituting (4.5) into (4.3) gives,

$$\frac{X^{IV}}{X} - 2\lambda^2 \frac{X''}{X} + \lambda^4 = 0$$

$$X^{IV} - 2\lambda^2 X'' + \lambda^4 x = 0$$

Let $X = e^{\alpha x}$

$$\alpha^4 - 2\lambda^2 \alpha^2 + \lambda^4 = 0 = (\alpha^2 - \lambda^2)^2$$

$$\text{or} \quad \alpha = \pm \lambda, \pm \lambda \quad \text{or} \quad X = \begin{Bmatrix} \cosh \lambda x \\ x \cosh \lambda x \\ \sinh \lambda x \\ x \sinh \lambda x \end{Bmatrix}.$$

$$(4.7)$$

So, there are eight such products as solutions of where

$$w(x, y) = X(x)Y(y) \quad \text{and} \quad k(y) = -\lambda^2.$$

Case 1b $\lambda^2 = 0$

$$k(y) = \frac{Y''}{Y} = 0 \quad Y = \begin{Bmatrix} 1 \\ y \end{Bmatrix}.$$

Substituting (4.5) into (4.3) gives

$$\frac{X^{iv}}{X} - 2\lambda^2 \frac{X''}{X} + \lambda^4 = 0$$

$$\left. \vphantom{\begin{matrix} 1 \\ 1 \\ 1 \\ 1 \end{matrix}} \right\} \quad (4.8)$$

If

$$\lambda^2 = 0$$

$$X = \begin{Bmatrix} 1 \\ x \\ x^2 \\ x^3 \end{Bmatrix}.$$

$$X^{iv} = 0$$

Hence, another eight products are found to be solutions to $\nabla^4 w = 0$ where $w = X(x)Y(y)$ and $k(y) = 0$.

Case 1c $\lambda^2 < 0$ (λ is imaginary). Hence, $\lambda = i\bar{\lambda}$

$$Y'' + \lambda^2 Y = 0 = Y'' - \bar{\lambda}^2 Y = 0 \quad Y = \begin{Bmatrix} \sinh \bar{\lambda}y \\ \cosh \bar{\lambda}y \end{Bmatrix},$$

as before $\alpha = \pm \lambda, \pm \lambda$ where λ is imaginary, let

$$X = e^{\alpha x} \quad \text{so}$$

$$\frac{X^{iv}}{X} = -2\lambda^2 \frac{X^4}{X} + \lambda^4 = 0$$

$$\left. \vphantom{\begin{matrix} 1 \\ 1 \\ 1 \end{matrix}} \right\} \quad (4.9)$$

$$\alpha^4 + 2\bar{\lambda}^2\alpha^2 + \bar{\lambda}^4 = 0 \quad X = \begin{Bmatrix} \cos \bar{\lambda}x \\ x \cos \bar{\lambda}x \\ \sin \bar{\lambda}x \\ x \sin \bar{\lambda}x \end{Bmatrix}.$$

$$\alpha = +i\bar{\lambda}, \pm i\bar{\lambda}$$

Case II

$$g(x) = -\gamma^2 \quad (4.10)$$

Case IIa $\gamma^2 > 0$

$$g(x) = \frac{X''}{X} = -\gamma^2$$

$$X'' + \gamma^2 X = 0 \quad X = \begin{Bmatrix} \cos \gamma x \\ \sin \gamma x \end{Bmatrix}$$

and as before

$$Y = \begin{Bmatrix} \cosh \gamma y \\ y \cosh \gamma y \\ \sinh \gamma y \\ y \sinh \gamma y \end{Bmatrix}$$

$$\left. \vphantom{\begin{array}{c} a \\ b \\ c \\ d \\ e \\ f \end{array}} \right\} \quad (4.11)$$

Case IIb $\gamma^2 = 0$

then

$$X = \begin{Bmatrix} 1 \\ X \end{Bmatrix}$$

$$Y = \begin{Bmatrix} 1 \\ y \\ y^2 \\ y^3 \end{Bmatrix}$$

$$\left. \vphantom{\begin{array}{c} a \\ b \\ c \\ d \\ e \\ f \end{array}} \right\} \quad (4.12)$$

Case IIc $\gamma^2 < 0$

let $\gamma = i\bar{\gamma}$

$$X = \begin{Bmatrix} \cosh \bar{\gamma} x \\ \sinh \bar{\gamma} x \end{Bmatrix}$$

$$Y = \begin{Bmatrix} \sin \bar{\gamma} y \\ y \sin \bar{\gamma} y \\ \cos \bar{\gamma} y \\ y \cos \bar{\gamma} y \end{Bmatrix}$$

$$\left. \vphantom{\begin{array}{c} a \\ b \\ c \\ d \\ e \\ f \end{array}} \right\} \quad (4.13)$$

So in each case there are eight possible products to satisfy any particular case. These solutions comprise all the possible separable solutions of the homogeneous biharmonic equation. In the solution of any particular problem one can attempt to find the solution through exploiting the particular boundary conditions and loading, intuition and experience. However if that fails then one can resort to trying each of the above solutions for the homogeneous solution.

4.2. Double Series Solution (Navier Solution)

In plate problems one can usually obtain solutions using a doubly infinite series, such as

$$w(x, y) = \sum_{m=1}^{\infty} \sum_{n=1}^{\infty} A_{mn} f_m(x) g_n(y) .$$

Such solutions are often inefficient to compute with due to the very slow convergence of the series. Instead one usually tries to obtain a solution where the function of only one spatial variable is summed such that, in this case:

$$w(x, y) = \sum_{n=1}^{\infty} \phi_n(y) f_n(x) .$$

This approach is particularly useful when two opposite edges are simply supported, because then the function $f_n(x)$ above can be a half range sine series. This is discussed in the next section.

In assuming the functions $f_m(x)$ and $g_n(y)$ for the double series solution (Navier Solution), or assuming the functions $f_n(x)$ for the single series solution (the M. Levy solution), the functions must be complete in order that the lateral deflection can be adequately represented. Furthermore, it is most convenient from a computational point of view that the functions be orthogonal. Also of course they must satisfy the boundary conditions for the problem. One straightforward approach to selecting such functions is to use the vibration modes or buckling modes for a beam of constant cross section with the same boundary conditions as those on opposite edges of the plate, because all such modes comprise a complete, orthogonal set. The beam vibration modes for all boundary conditions and their properties have been conveniently catalogued by Young and Felgar (Reference 4.1) and Felgar (Reference 4.2).

The doubly infinite series approach will be treated first. Consider a rectangular plate simply supported on all four edges in the region $0 \leqslant x \leqslant a, \ 0 \leqslant y \leqslant b, \ -h/2 \leqslant z \leqslant h/2$.

The governing equation is:

$$\nabla^4 w = p(x, y)/D$$

The solution can be written as

$$\text{Let } w(x, y) = \sum_{m=1}^{\infty} \sum_{n=1}^{\infty} A_{mn} \sin \frac{m \pi x}{a} \sin \frac{n \pi y}{b} , \tag{4.14}$$

because these functions are complete, orthogonal and they satisfy the boundary conditions of the problem. The lateral load must be expanded in the same series solution:

$$p(x, y) = \sum_{m=1}^{\infty} \sum_{n=1}^{\infty} B_{mn} \sin \frac{m \pi x}{a} \sin \frac{n \pi y}{b} \tag{4.15}$$

where

$$B_{mn} = \frac{4}{ab} \int_0^a \int_0^b p(x, y) \sin \frac{m \pi x}{a} \sin \frac{n \pi y}{b} \, dy \, dx . \tag{4.16}$$

Substituting these series representations of the load and lateral deflection into the governing differential equation produces

$$\sum \sum A_{mn} \pi^4 \left\{ \frac{m^4}{a^4} + 2 \frac{m^2}{a^2} \frac{n^2}{b^2} + \frac{n^4}{b^4} \right\} \sin \frac{m\pi x}{a} \sin \frac{n\pi y}{b}$$

$$= \frac{1}{D} \sum \sum B_{mn} \sin \frac{m\pi x}{a} \sin \frac{n\pi y}{b} .$$

For the doubly infinite series above to be equal requires that an equality exist for each m and n combination in the series. Looking at the mth and nth term, A_{mn} is easily found in terms of B_{mn}.

$$A_{mn} = \frac{B_{mn}}{D\pi^4 \left\{ \frac{m^2}{a^2} + \frac{n^2}{b^2} \right\}^2} . \tag{4.17}$$

Thus, the solution is easily found for this case, because B_{mn} is determined from (4.16), and A_{mn} is then found from the equation above, hence $w(x, y)$ is then known everywhere from (4.14). From this, all slopes, stress couples, and shear resultants can be calculated at any location x, y. As mentioned previously, the doubly infinite series solution usually converges slowly. Moreover, the derivatives of $w(x, y)$ needed to obtain stress couples and shear resultants always converge still slower than the deflection function itself.

An example of obtaining B_{mn} can be briefly given. Consider a plate simply supported on all four edges subjected to a uniform constant lateral loading p_0.

$$B_{mn} = \frac{4p_0}{ab} \int_0^a \int_0^b \sin \frac{m\pi x}{a} \sin \frac{n\pi y}{b} \, dy \, dx$$

$$= \frac{4p_0}{ab} \left\{ \frac{a}{m\pi} \left[-\cos \frac{m\pi x}{a} \right]_0^a \right\} \left\{ \frac{b}{n\pi} \left[-\cos \frac{n\pi y}{b} \right]_0^b \right\}$$

$$= \frac{4p_p}{mn\pi^2} (1 - \cos m\pi)(1 - \cos n\pi)$$

$$= \frac{4p_0}{mn\pi^2} [1 - (-1)^m][1 - (-1)^n)$$

$$= \frac{16p_0}{mn\pi^2} \quad \text{(if, } m, n \text{ odd only)} \tag{4.18}$$

4.3. Single Series Solution (Method of M. Levy)

Consider a plate with opposite edges simply supported, as shown in Figure 4.1.

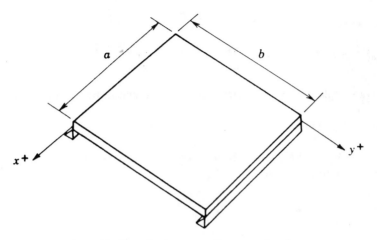

Figure 4.1. Plate simply supported on opposite edges.

Again, the governing differential equation is:

$$\nabla^4 w = \frac{p(x, y)}{D}.$$

The boundary conditions on the y edges are:

$$w(x, 0) = w(x, b) = 0$$
$$M_y(x, 0) = M_y(x, b) = 0.$$

(4.19)

From (2.38), the stress couple is given by

$$M_y = -D\left[\frac{\partial^2 w}{\partial y^2} + v \frac{\partial^2 w}{\partial x^2}\right].$$

Hence on the $y = 0$ and $y = b$ edges,

$$\frac{\partial^2 w}{\partial y^2}\left(x, \frac{0}{b}\right) + v \frac{\partial^2 w}{\partial x^2}\left(x, \frac{0}{b}\right) = 0.$$

However,

$$\frac{\partial^2 w}{\partial x^2}\left(x, \frac{0}{b}\right) = 0$$

because the curvature is zero parallel to the simply supported edge. Therefore,

$$\frac{\partial^2 w}{\partial y^2}\left(x, \frac{0}{b}\right) = 0.$$

(4.20)

Assume a form of the solution to be as follows, which satisfies the boundary condition on the y edges given by (4.19):

$$w(x, y) = \sum_{n=1}^{\infty} \phi_n(x) \sin \frac{n\pi y}{b} . \tag{4.21}$$

For this example, the lateral distributed load is taken to be the following:

$$p(x, y) = g(x)h(y) \tag{4.22}$$

where $g(x)$ and $h(y)$ are given. It is necessary to expand $h(y)$ in a series solution corresponding to (4.21), hence,

$$h(y) = \sum_{n=1}^{\infty} A_n \sin \frac{n\pi y}{b}$$

where   (4.23)

$$A_n = \frac{2}{b} \int_0^b h(y) \sin \frac{n\pi y}{b} \, dy .$$

Substituting (4.21) through (4.23) into (4.1) gives:

$$\sum_{n=1}^{\infty} \{\phi_n^{iv} - 2\lambda_n^2 \phi_n'' + \lambda_n^4 \phi_n\} \sin \frac{n\pi y}{b} = \frac{1}{D} \sum_{n=1}^{\infty} A_n g_n(x) \sin \frac{n\pi y}{b} \tag{4.24}$$

where $\lambda_n = n\pi/b$.

For this to be true, the series must be equated term by term.

$$\phi_n^{iv}(x) - 2\lambda_n^2 \phi_n''(x) + \lambda_n^4 \phi_n(x) = \frac{1}{D} A_n g_n(x) . \tag{4.25}$$

Note, at this point the boundary conditions on the other two edges have not been specified. Thus, any time a problem has two opposite edges simply supported, one can arrive at (4.25) without any other information.

Proceeding to solve (4.25) in the customary way, let $\phi_n = e^{sx}$ such that the homogeneous solution becomes:

$$s^4 - 2\lambda_n^2 s^2 + \lambda_n^4 = 0$$

$$(s^2 - \lambda_n^2)(s^2 - \lambda_n^2) = 0 \quad \text{where } \lambda_n^2 > 0$$

$$s = \pm \lambda_n, \ \pm \lambda_n$$

So, the complementary solution is:

$$\phi_n(x) = (C_1 + C_2 x) \cosh \lambda_n x + (C_3 + C_4 x) \sinh \lambda_n x . \tag{4.26}$$

Equation (4.26) is the form of the homogeneous solution for $\lambda_n(x)$ for any set of boundary conditions on the x-edges.

4.3.1. Example: Plate Simply Supported on All Four Edges and $p = p(y)$ Only

On the $x = $ constant edges, the boundary conditions are:

$$w(0, y) = w(a, y) = 0$$

$$M_x(0, y) = M_x(a, y) = 0 .$$

(4.27)

Because of no curvature along the $x = $ constant edges, i.e.,

$$\frac{\partial^2 w}{\partial y^2}\left(\frac{0}{a}, y\right) = 0$$

the moment boundary conditions can be written as:

$$\frac{\partial^2 w}{\partial x^2}(0, y) = \frac{\partial^2 w}{\partial x^2}(a, y) = 0 .$$

(4.28)

Also since $p = p(y)$ in this example, $g(x) = 1$ in (4.22), and from (4.25) the particular solution can be written as:

$$\phi_{n_p}(x) = \frac{A_n}{D\lambda_n^4} .$$

(4.29)

Therefore for this example, the complete solution for $\phi_n(x)$ is:

$$\phi_n(x) = (C_1 + C_2 x)\cosh \lambda_n x + (C_3 + C_4 x)\sinh \lambda_n x + \frac{A_n}{D\lambda_n^4} .$$

(4.30)

Substituting (4.30), the complete solution for $\phi_n(x)$, and its derivatives into (4.27) and (4.28), the boundary conditions on the x edges, provides the values of the undetermined constants C_1 through C_4, for this problem. The results are:

$$C_1 = -\frac{A_n}{D\lambda_n^4}$$

$$C_2 = \frac{A_n}{2D\lambda_n^3}\frac{1 - \cosh \lambda_n a}{\sinh \lambda_n a}$$

$$C_3 = \frac{A_n a}{2D\lambda_n^3(1 + \cosh \lambda_n a)}\left[\frac{2}{\lambda_n a}\sinh \lambda_n a - 1\right]$$

$$C_4 = \frac{A_n}{2D\lambda_n^3}$$

(4.31)

Thus, the complete solution for the lateral deflection is:

$$w(x, y) = \sum_{n=1}^{\infty}\left[(C_1 + C_2 x)\cosh \lambda_n x + (C_3 + C_4 x)\sinh \lambda_n x + \frac{A_n}{D\lambda_n^4}\right]\sin \lambda_n y$$

(4.32)

where C_1 through C are given by (4.31).

It should be noted that we could have solved this problem by assuming the deflection any one of these following ways

$$w(x, y) = \sum_{m=1}^{\infty} \sum_{n=1}^{\infty} A_{mn} \sin \frac{m\pi x}{a} \sin \frac{n\pi y}{b}$$

$$w(x, y) = \sum_{n=1}^{\infty} \phi_n(x) \sin \frac{n\pi y}{b}$$

$$w(x, y) = \sum_{m=1}^{\infty} \psi_m(y) \sin \frac{m\pi x}{a} .$$

The first of these methods converges more slowly, while the second and third converge quite rapidly.

For the case of the x edges being clamped or free, and with the same loading, $p = p(y)$ only, Equation (4.30) with the appropriate boundary conditions to obtain the solution may be used.

4.4. Example of Plate with Edges Supported by Beams

The use of beams to support plate elements is very commonplace. Innovative and efficient design often results in complex analytical procedures, so complicated in fact that doctoral dissertations have been written in this regard. Complications can arise when the plate mid-surface is not also the mid-surface of the support beams, beam sections involving centers of twist in difficult locations, discontinuous joining of beam and plate, etc.

Presented here is the simplest of beam-plate combinations, merely to introduce the concepts involved.

Consider a rectangular plate with the following boundary conditions: $y = 0, b$ simply supported; $x = 0, a$ supported by beams; and a lateral load given by $p(x, y) = \sum_{n=1}^{\infty} A_n \sin(n\pi y/b)$.

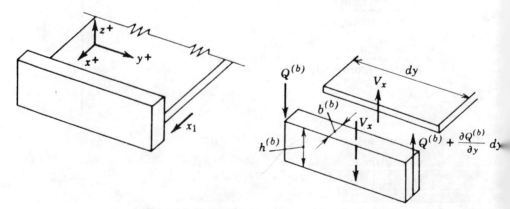

Figure 4.2. Vertical forces at beam supported edge.

For one boundary condition on the x edge; consider an element of beam as a free body, as shown in Figure 4.2. A force balance in the z direction provides one plate boundary condition.

Looking at the details of the Figure 4.2, the superscripted b quantities refer to a beam, whose flexural stiffness is $(EI)^b$, which is mechanically joined to the edge of the plate denoted by $x = x_1$ such that the middle surface of both the plate and beam are identical, to retain simplicity in this example. Hence, the lateral deflection of the beam and plate are identical at their common boundary, $x = x_1$. Therefore, the force balance is given by the following, where the shear resultant of the beam is

$$Q^{(b)}(y) = b^{(b)} \int_{-h^{(b)}/2}^{+h^{(b)}/2} \sigma_{yz}\, dz$$

and the Kirchoff 'effective' shear resultant is used for the plate.

$$\sum F_z^{(b)} = 0 = -Q^{(b)} - V_x\, dy + Q_b^{(b)} + \frac{dQ^{(b)}}{dy}\, dy = 0$$

$$(V_x)_{x=x_1} = \frac{dQ^{(b)}}{dy}(x_1, y) = -(EI)^{(b)}\left(\frac{\partial^4 w^{(b)}}{\partial y^4}\right)_{x=x_1} = -(EI)^b\left(\frac{\partial^4 w}{\partial y^4}\right)_{x=x_1}$$

since $w_b = w$ at $x = x_1$ and since $Q^{(b)} = dM^{(b)}/dy$, and $M^{(b)} = -(EI)^{(b)}\, \partial^2 w^{(b)}/\partial y^2$.
  For the plate

$$(V_x)_{x=x_1} = -D\left[\frac{\partial^3 w}{\partial x^3} + (2 - v)\frac{\partial^3 w}{\partial x\, \partial y^2}\right]_{x=x_1}.$$

The second boundary condition, the balancing of twisting moments, provides the requirement. The beam has a torsional stiffness $(GJ)^b$ (Figure 4.3).
For beam, at $x = x_1$

$$-T - M_x\, dy + T + \frac{dT}{dy}\, dy = 0$$

$$M_x = \frac{dT}{dy} = -(GJ)^{(b)}\frac{d^2\theta^{(b)}}{dy^2} = -(GJ)^{(b)}\left(\frac{\partial^3 w}{\partial x\, \partial y^2}\right)$$

$$= -D\left[\frac{\partial^2 w}{\partial x^2} + v\frac{\partial^2 w}{\partial y^2}\right]$$

since

$$T = -(GJ)^{(b)}\frac{d\theta^{(b)}}{dy} \quad \text{and} \quad \theta^{(b)} = \frac{\partial w^{(b)}}{\partial x}$$

and

$$\frac{dw^{(b)}}{dx} = \frac{dw}{dx} \quad \text{at} \quad x = x_1.$$

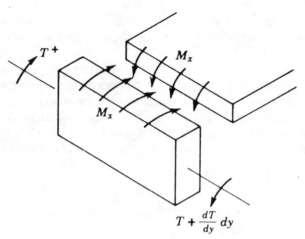

Figure 4.3. Stress couples at beam supported edges.

So the boundary conditions are:

$$D\left[\frac{\partial^3 w}{\partial x^3} + (2 - v)\frac{\partial^3 w}{\partial x\,\partial y^2}\right] = (EI)^{(b)}\frac{\partial^4 w}{\partial y^4} \quad \text{at} \quad x = x_1 \tag{4.33}$$

$$D\left[\frac{\partial^2 w}{\partial x^2} + v\frac{\partial^2 w}{\partial y^2}\right] = (GJ)^{(b)}\frac{\partial^3 w}{\partial x\,\partial y^2} \quad \text{at} \quad x = x_1 . \tag{4.34}$$

Once the plate deflection at $x = x_1$ is known, then one knows the deflection of the beam, and so the complete solution of the beam is determined also.

For all practical cases one can assume that the beam end conditions are the same as the plate end conditions, and assume that the ends of the beam is completely restrained against rotation.

4.5. Summary

In this chapter the two basic approaches to solving problems of rectangular plates subjected to lateral loads has been treated. Also, a more complicated boundary condition example was investigated than the classical boundary conditions of Chapter 1. In rectangular plates with more difficult boundary conditions than simply supported edges References 4.1 and 4.2 provide functions suitable for either the Navier or the Levy Method.

Many solutions to plate problems are known, and are catalogues in numerous references such as Timoshenko and Woinowsky-Krieger (Reference 4.3), Marguerre and Woernle (Reference 4.4) and Mansfield (Reference 4.5).

Methods of analysis are developed and example solutions are given for rectangular plates of anisotropic materials and composite materials in a texts by Vinson and Chou (Reference 4.6), and Vinson and Sierakowski (Reference 4.7), as well as several other texts.

4.6. References

4.1. Young, D. and R. P. Felgar, Jr., *Tables of Characteristic Functions Representing Normal Modes of Vibration of a Beam*, The University of Texas Engineering Research Series Report No. 44, July 1, 1949.

4.2. Felgar, R. P. Jr., *Formulas for Integrals Containing Characteristic Functions of a Vibrating Beam*, The University of Texas Bureau of Engineering Research Circular No. 14, 1950.

4.3. Timoshenko, S. and A. Woinowsky-Krieger, *Theory of Plates and Shells*, New York: McGraw-Hill Book Co., Inc., 2nd Edition, 1959.

4.4. Marguerre, K. and H. T. Woernle, *Elastic Plates*, Blaisdell Publishing Company, 1970.

4.5. Mansfield, E. H., *The Bending and Stretching of Plates*, Pergamon Press, 1964.

4.6. Vinson, J. R. and T. W. Chou, *Composite Materials and Their Use in Structures*, London: Applied Science Publishers, Ltd., 1974.

4.7. Vinson, J. R. and R. L. Sierakowski, *The Behavior of Structures Composed of Composite Materials*, Dordrecht: Martinus-Nijhoff Publishers, 1986.

4.7. Problems

4.1. Consider a rectangular isotropic plate occupying the region $0 \leqslant x \leqslant a$, $0 \leqslant y \leqslant b$, and $-h/2 \leqslant z \leqslant h/2$. The plate is simply supported on the edges $y = 0$ and b. The plate is subjected to a laterally distributed load given by Equations (4.22) and (4.23). If $g(x) = 1$, the solution is given by Equation (4.32). In the plate clamped along the edges $x = 0$ and a, determine C_1 through C_4.

4.2. In problem 4.1 above, if the plate is free along the edges $x = 0$ and a, determine the constants C_1 through C_4.

4.3. In problem 4.1 above, if the plate is simply supported at $x = 0$ and clamped at $x = a$, determine the constants C_1 through C_4.

4.4. In problem 4.1 above, if the plate is simply supported at $x = 0$ and free along $x = a$, determine the constants C_1 through C_4.

4.5. In problem 4.1 above, for the plate clamped along $x = 0$ and free along $x = a$ determine the constants C_1 through C_4.

4.6. Consider a floor slab whose geometry is described in problem 4.1. The slab is square, simply supported on all edges, and is loaded with sand in such a way that the load can be approximated by

$$p(x, y) = p_0 \sin \frac{\pi x}{a} \sin \frac{\pi y}{b} .$$

Determine the location and magnitude of the maximum deflection, the maximum bending stresses in both directions, and the maximum shear stresses in each direction.

4.7. A certain window in an aircraft is approximated by a square plate of dimension a on each side, simply supported on all four edges and subjected to a uniform cabin pressure p_0. Using the Navier solution for a square plate of length and width a, the solution is given by Equation (4.14). The maximum value of the lateral deflection can be written as

$$W_{\max} = C_1 p_0 a^4 / D$$

for the plate subjected to a constant lateral loading, p_0. Determine the numerical coefficient C_1 to three significant figures.

The maximum bending moment $M_{x_{\max}} = M_{y_{\max}}$ can be written as

$$M_{\max} + C_2 p_0 a^2 .$$

Find C_2 to three significant figures, if the Poisson's ratio of the window material is $v = 0.3$.

4.8. A certain hull plate on the flat bottom of a ship may be considered to be a rectangular plate under uniform loading, p_0, from the water pressure, and clamped along all edges. A 1/2″ steel plate four feet in width is to be used for the bottom plate in the ship draws 13 1/2 feet of water maximum. If the maximum allowable stress in the steel is 20 000 psi, what is the maximum plate length (i.e., bulkhead spacing that can be used in the ship design, and what is the corresponding maximum deflection of the hull plate. Salt water weighs 64 lbs/ft³, $E_{steel} = 30 \times 10^6$ psi, and $v_{steel} = 0.3$.

For a plate clamped on all four edges, subjected to a lateral load p_0, the maximum deflection and maximum stress couple can be written as follows, and the value of C_1 and C_2 calculated for various plate aspect ratios.

$$w_{max} = C_1 p_0 a^4 / E h^3, \quad \text{where } v = 0.3$$

$$M_{max} = M_{x_{max}} = C_2 p_0 a^2, \quad \text{where } v = 0.3$$

b/a	1	1.2	1.4	1.6	1.8	2.0	3.0
C_1	0.0138	0.0188	0.0226	0.0251	0.0267	0.0277	0.0285
C_2	0.0513	0.0639	0.0726	0.0780	0.0812	0.0829	0.0833

Linear interpolation is permitted

4.9. A rectangular wing panel component, $8'' \times 5''$ is made of aluminum, and under the most severe maneuver conditions can be subjected to a uniform lateral load of 20 psi. This wing panel can be approximated by a flat plate simply supported on all four edges. What thickness must the panel be, and what is the resulting maximum deflection under this maneuver condition?

A rectangular plate simply supported on all four edges subjected to a uniform lateral load, with $v = 0.3$, the maximum deflection w_{max} and the maximum moment $M_x = M_{max}$, located at $(x = a/2, y = b/2)$, can be expressed as

$$w_{max} = C_1 p_0 a^4 / E h^3$$

$$M_{max} = M_{x_{max}} = C_2 p_0 a^2.$$

b/a	1	1.2	1.4	1.6	1.8	2.0	3.0	4.0	5.0	∞
C_1	0.044	0.062	0.077	0.091	0.102	0.111	0.134	0.140	0.142	0.142
C_2	0.042	0.063	0.075	0.086	0.095	0.102	0.119	0.124	0.125	0.125

The aluminum used has an allowable stress of 20 000 psi, and $E = 10 \times 10^6$ psi and $v = 0.3$.

4.10. A rectangular steel plate is used as part of a flood control structure, and is mounted vertically under water such that it is subjected to a hydraulic loading

$$p(x, y) = p_0 + p_1 \frac{y}{b}$$

where p_0 and p_1 are constants associated with the pressure heads. Find the Euler coefficient B_{mn} for this loading in Equation (4.15).

4.11. A glass manufacturer has been asked to construct plate glass windows for a new modern office building. The windows must be 10 ft wide and 20 ft high. Design the windows so that they can withstand wind forces due to air velocities of 150 miles/hour. State all assumptions and physical constants clearly.

4.12. A flat portion of a wind tunnel measuring $30'' \times 54''$ will be subjected to a maximum uniform wind load of 10 psi. If the steel to be used has an allowable stress of 40 000 psi, and a Poisson's ratio of $v = 0.3$, what plate thickness is required if the plate is
(a) Simply supported on all four edges.
(b) Clamped on all four edges.
Use the data from problems 4.8 and 4.9.

4.13. A portion of the cover on a hover craft is to be rectangular measuring 8′ × 4′ in planform, and is to be simply supported on all four edges. It is calculated that the maximum air pressure the panel will be subjected to is 20 psi.
   (1) How thick must the panel be if it is constructed of aluminum ($E = 10 \times 10^6$ psi, $v = 0.3$) if the allowable stress is limited to 30 000 psi?
   (2) How thick must the plate be if it is constructed of steel ($E = 30 \times 10^6$ psi, $v = 0.3$) if the allowable stress is limited to 60 000 psi?
   (3) If the weight density of steel is 0.283 lbs/in³ and that of aluminum is 0.1 lbs/in³, which material should be selected to minimize weight?
   (4) Suppose the aluminum plate of (1) above were clamped on all four edges, what thickness is required?
   Use the data of Problems 4.8 and 4.9.

4.14. A rectangular steel plate, used as a footing, rests on the ground is subjected to a uniform lateral pressure, $p(x, y) = -P_0$ (psi). The ground deflects linearly below the footing with a spring constant k (lbs/in²/in) under this loading and deflection.

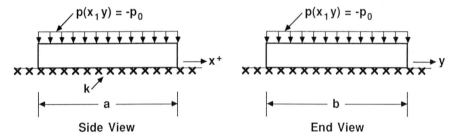

Figure 4.4. Elasticically supported footing.

   a. What is the governing differential for the bending of this plate?
   b. Where are the boundary conditions on the $x = 0$ and $x = a$ edges?
   c. What are the boundary conditions on the $y = 0$ and $y = b$ edges?

4.15. A designer is faced with the problem of designing a rectangular plate cover over an opening that is 9 feet by 3 feet. The design load is a lateral pressure of 10 psi. If steel is used ($E = 30 \times 10^6$ psi, $v = 0.3$, σ allowable = 35 000 psi, $\rho = 0.283$ lbs/in³):
   a. If the plate is clamped on all four edges, what will it weigh?
   b. If the plate is simply supported on all four edges, what will it weigh?

4.16. (a) A designer is considering two alternatives in designing a flat plate made of aluminum, which measures 60″ × 30″. It must withstand an in-plane compressive load in the longer direction of $N_x = -100$ lbs/in. To simply support the plate on all four edges requires 20 lbs of support structure; to clamp the plate on all four edges requires 40 lbs of support structure. The plate structures must not be overstressed. Which design will result in the lighter overall structural system: for aluminum: $E = 10 \times 10^6$ psi, $v = 0.3$, allowable = 30000 psi, $\rho = 0.1$ lbs/in³.
   (b) The designer could also choose to use magnesium with $E = 6.5 \times 10^6$ psi, $v = 0.3$, allowable -30000 psi, $p = 0.1$ lbs/in³. If simply supported on all four edges, would a magnesium overall plate structural system be lighter than the simply supported aluminum system?
   (c) If the designer used steel with $E = 30 \times 10^6$ psi, $v = 0.3$, allowable = 60000 psi and $\rho = 0.283$ lb/in³, would the overall structural system, simply supported, be lighter than the above?

4.17. A rectangular aluminum plate, measuring 40″ × 20″, is subjected to a uniform lateral pressure of 10 psi. Using the maximum stress theory, if the allowable stress is 30000 psi, what is the plate thickness required if:
   a. All edges of the plate are simply supported?
   b. All edges of the plate are clamped?

    c. If the plate were made of steel with the same allowable stress as the aluminum above, would the required thickness differ from that of the aluminum plate?

    d. If the plate were made of steel with the same allowable stress as the aluminum above, would the maxium deflection differ from that of the aluminum plate?

4.18. Consider a plate clamped on all four edges made of the same steel as in Problem 3.1. The plate is subjected to a uniform lateral load of $p = 10$ psi. If the plate is $10''$ wide and $16''$ long, what thickness h is required to prevent overstressing or a maximum deflection of $0.1''$?

5

Thermal Stresses in Plates

5.1. General Considerations

Consider any elastic body with a constant coefficient of thermal expansion, α, in the units of $in/in/°F$, or equivalent, at a uniform temperature wherein the body is assumed to be free of thermal stresses and strains. If the body is free to deform, and the temperature is raised slowly to a temperature of T degrees from the stress free temperature, the thermal strains produced at any material point can be written as

$$\varepsilon_{ij_{th}} = \alpha\Delta T(x_i)\delta_{ij} \tag{5.1}$$

where x_1 are the coordinate directions, and δ_{ij} is the Kronecker delta ($\delta_{ij} = 1$ for $i = j$, $\delta_{ij} = 0$ for $i \neq j$). It should be noted that thermal strains are purely dilatational ($i = j$); thermal shear strains do not exist.

In Equation (5.1), ΔT is positive when the temperature of the material point is above the stress free temperature. The coefficient of thermal expansion α is positive for all isotropic engineering materials, i.e., the body expands when it is heated.

In many thermoelastic bodies, the changes in temperature within the body tend to result in strains which do not satisfy the compatibility equations. In that case isothermal strains, $\varepsilon_{ij_{iso}}$, the strains discussed in Chapter 1, are induced such that the total strain, $\varepsilon_{ij_{tot}}$, satisfies compatibility.

$$\varepsilon_{ij_{tot}} = \varepsilon_{ij_{iso}} + \varepsilon_{ij_{th}} . \tag{5.2}$$

In that case the 'thermal stresses' are induced due to the isothermal strains induced to insure compatibility. This can occur, for example, in 'thermal shock' from very rapid or localized heating.

A second way that thermal stresses occur is through displacement restrictions on the elastic body. One simple example of this occurs when a bar is placed between immovable end grips and subsequently heated. There, compressive thermal stresses result.

Hence, thermal stresses are caused by two mechanisms: one by displacement restrictions, the other through induced isothermal strains to maintain compatibility.

Next, consider an unrestricted thin rod at a uniform temperature. If the rod is slowly heated uniformly, such that the thermal strains satisfy the compatibility equations, the heated rod has thermal strains but no thermal stresses. Now if the unheated thin rod is placed in immovable end grips such that the rod cannot increase in length, slowly heating the rod uniformly will result in thermal stresses and no thermal strains.

In the latter case, if the compressive axial thermal stresses reach a value equal to the Euler buckling load (discussed later in Chapter 7), the rod will buckle. This is called thermal buckling.

5.2. Derivation of the Governing Equations for a Thermoelastic Plate

In deriving the governing equations for a thermoelastic plate, the equilibrium equations and the strain-displacement equations are not altered from those of the isothermal plate of Chapter 1, because in the former the equations involve force balances, and the latter are purely kinematic relationships involving total strains.

However, the stress strain relations, Equations (1.9) and (1.10), are modified in accordance with Equations (5.2) and (5.1):

$$\varepsilon_{x_{\text{iso}}} = \varepsilon_{x_{\text{tot}}} - \alpha \Delta T = \frac{1}{E} \left[\sigma_x - v\sigma_y \right]$$

$$\varepsilon_{y_{\text{iso}}} = \varepsilon_{y_{\text{tot}}} - \alpha \Delta T = \frac{1}{E} \left[\sigma_y - v\sigma_x \right]$$

or

$$\varepsilon_x = \frac{1}{E} \left[\sigma_x - v\sigma_y \right] + \alpha \Delta T \tag{5.3}$$

$$\varepsilon_y = \frac{1}{E} \left[\sigma_y - v\sigma_x \right] + \alpha \Delta T. \tag{5.4}$$

In Equations (5.3) and (5.4) and in all that follows the subscript for total strains is dropped, and all strains noted explicitly are those which satisfy compatibility, and which appear in the strain-displacement relations. Hence, in Equations (5.3) and (5.4) the first terms on the right-hand side are really the isothermal strains, and the second terms on the right-hand side are thermal strains.

Proceeding as in Chapter 1, employing the strain displacement relations, Equations (1.16) and (1.17), Equations (5.3) and (5.4) after multiplying by E become

$$\sigma_x - v\sigma_y + E\alpha \Delta T = E \frac{\partial u_0}{\partial x} - Ez \frac{\partial^2 w}{\partial x^2} \tag{5.5}$$

$$\sigma_y - v\sigma_x + E\alpha \Delta T = E \frac{\partial v_0}{\partial y} - Ez \frac{\partial^2 w}{\partial y^2} \tag{5.6}$$

Now, two quantities N^* and M^*, known as the thermal stress resultant and the thermal stress couple, respectively, are defined as

$$N^* = \int_{-h/2}^{+h/2} E\alpha\Delta T\,dz\,, \quad M^* = \int_{-h/2}^{+h/2} E\alpha\Delta Tz\,dz\,. \tag{5.7}$$

Multiplying Equations (5.5) and (5.6) by dz and integrating across the thickness of the plate, then multiplying them by $z\,dz$ and analogous by integrating them, provides the integrated stress strain relations for a thermoelastic plate. It should be remembered from the discussion of Section 5.1 that the shear stress-strain relations are not altered by the inclusion of thermoelastic effects.

$$N_x - \nu N_y + N^* = Eh\,\frac{\partial u_0}{\partial x}$$

$$N_y - \nu N_x + N^* = Eh\,\frac{\partial v_0}{\partial y}$$

$$M_x - \nu M_y + M^* = \frac{-Eh^3}{12}\,\frac{\partial^2 w}{\partial x^2}$$

$$M_y - \nu M_x + M^* = \frac{-Eh^3}{12}\,\frac{\partial^2 w}{\partial y^2}$$

$$N_{xy} = \frac{K(1-\nu)}{2}\left[\frac{\partial u_0}{\partial y} + \frac{\partial v_0}{\partial x}\right] \tag{5.8}$$

$$M_{xy} = -D(1-\nu)\,\frac{\partial^2 w}{\partial x\,\partial y}\,. \tag{5.9}$$

Rearranging the first four of the above results in

$$N_x = K\left[\frac{\partial u_0}{\partial x} + \nu\,\frac{\partial v_0}{\partial y}\right] - \frac{N^*}{(1-\nu)} \tag{5.10}$$

$$N_y = K\left[\frac{\partial v_0}{\partial y} + \nu\,\frac{\partial u_0}{\partial x}\right] - \frac{N^*}{(1-\nu)} \tag{5.11}$$

$$M_x = -D\left[\frac{\partial^2 w}{\partial x^2} + \nu\,\frac{\partial^2 w}{\partial y^2}\right] - \frac{M^*}{(1-\nu)} \tag{5.12}$$

$$M_y = -D\left[\frac{\partial^2 w}{\partial y^2} + \nu\,\frac{\partial^2 w}{\partial x^2}\right] - \frac{M^*}{(1-\nu)}\,. \tag{5.13}$$

Introducing these thermoelastic stress-strain relations into the equilibrium equations (2.44) through (2.46) and (2.50) and (2.51), the governing differential equations for a thermoelastic plate are determined, for the case of no surface shear stresses:

$$DV^4w = p(x, y) - \frac{1}{(1 - v)} \nabla^2 M^* \tag{5.14}$$

$$KV^4u_0 = \frac{1}{(1 - v)} \frac{\partial}{\partial x} (\nabla^2 N^*) \tag{5.15}$$

$$KV^4v_0 = \frac{1}{(1 - v)} \frac{\partial}{\partial y} (\nabla^2 N^*). \tag{5.16}$$

Also for completeness, other useful relationships are catalogued below:

$$Q_x = D \frac{\partial}{\partial x} (\nabla^2 w) - \frac{1}{(1 - v)} \frac{\partial M^*}{\partial x}$$

$$Q_y = -D \frac{\partial}{\partial y} (\nabla^2 w) - \frac{1}{(1 - v)} \frac{\partial M^*}{\partial y} \tag{5.17}$$

$$V_x = -D \left[\frac{\partial^3 w}{\partial x^3} + (2 - v) \frac{\partial^3 w}{\partial x \, \partial y^2} \right] - \frac{1}{(1 - v)} \frac{\partial M^*}{\partial x}$$

$$V_y = -D \left[\frac{\partial^3 w}{\partial y^3} + (2 - v) \frac{\partial^3 w}{\partial y \, \partial x^2} \right] - \frac{1}{(1 - v)} \frac{\partial M^*}{\partial y} \tag{5.18}$$

Due to the inclusion of thermal quantities, the expressions for various normal stresses in the plate are modified as follows:

$$\sigma_x = \frac{1}{h} \left[N_x + \frac{N^*}{(1 - v)} \right] + \frac{z}{h^3/12} \left[M_x + \frac{M^*}{(1 - v)} \right] - \frac{E \alpha \Delta T}{(1 - v)}$$

$$\sigma_y = \frac{1}{h} \left[N_y + \frac{N^*}{(1 - v)} \right] + \frac{z}{h^3/12} \left[M_y + \frac{M^*}{(1 - v)} \right] - \frac{E \alpha \Delta T}{(1 - v)}. \tag{5.19}$$

The inclusion of the N^* and M^* terms in (5.19) are easy to visualize since they are thermal stress resultants and couples analogous to N_x, N_y, M_x, and M_y, which are caused by lateral and in-plane 'mechanical' loads. The last terms in (5.19) can be visualized by the following example in which it is assumed that the first two terms do not contribute. Suppose at some value of (x, y) in a plate, the upper surface is heated while the lower surface is cooled. Thus the value of T in the upper portion of the plate is positive and becomes negative in the lower plate portion, as shown in the sketch below.

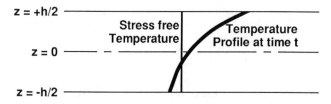

The last term of (5.19) shows that in the upper portion of the plate compressive stresses exist while in the lower portion there are tensile stresses due to the last term of (5.19). Physically, the material points in the upper portion of the plate want to expand considerably but are being restrained by those in the cooler areas of the plate, hence, tending to cause high compressive stresses there. Likewise, in the cooler portion of the plate the material points wish to contract, but are being extended by the hotter portions of the plate, hence, thrown into tension. Such thermal stresses can result in material failure just as stresses caused by mechanical loads.

As discussed before, shear stresses and strains are not affected by thermal effects, hence remain the same as in Chapter 2:

$$\sigma_{xy} = \frac{N_{xy}}{h} + \frac{M_{xy}z}{h^3/12}$$

$$\sigma_{xz} = \frac{3Q_x}{2h}\left[1 - \left(\frac{z}{h/2}\right)^2\right] \tag{5.20}$$

$$\sigma_{yz} = \frac{3Q_y}{2h}\left[1 - \left(\frac{z}{h/2}\right)^2\right].$$

Of course if there exist shear stresses applied to the upper or lower surfaces of the plate, the latter two expressions must be modified as in Chapter 2.

To proceed with solutions of thermoelastic plates using equations (5.14) through (5.16), one now proceeds using the same techniques that were introduced in Chapter 4. However, the additions of thermal effects do introduce certain difficulties with boundary conditions that cause some analytical difficulties. These are discussed in the next section.

5.3. Boundary Conditions

Looking now at the boundary conditions associated with a thermoelastic plate, comparisons are made with an isothermal plate, where again n denotes normal to the edge and s denotes along the edge:

Simply Supported Edge

$$w = 0$$
$$M_n = 0. \tag{5.21}$$

From Equations (5.12) and (5.13), the latter equation above is in fact

$$M_n = 0 = -D\left[\frac{\partial^2 w}{\partial n^2} + v\,\frac{\partial^2 w}{\partial s^2}\right] - \frac{M^*}{(1-v)}\,.$$

Since there is no curvature along the edge (i.e., $\partial^2 w/\partial s^2 = 0$), this equation becomes

$$\frac{\partial^2 w}{\partial n^2} = -\frac{M^*}{D(1-v)}\,. \tag{5.22}$$

Hence, the boundary conditions for a simply supported thermoelastic plate are nonhomogeneous.

Clamped Edge

$$w = 0 \tag{5.23}$$

$$\frac{\partial w}{\partial n} = 0\,. \tag{5.24}$$

These remain the same as those for the isothermal plate.

Free Edge

The boundary conditions are

$$M_n = 0 \quad \text{and} \quad V_n = 0\,.$$

Hence, the first condition is given by Equation (5.22) and the latter is seen to be from Equation (5.18).

$$\frac{\partial^3 w}{\partial n^3} + (2-v)\,\frac{\partial^3 w}{\partial n\,\partial s^2} = -\frac{1}{D(1-v)}\,\frac{\partial M^*}{\partial n}\,. \tag{5.25}$$

Here the boundary conditions are seem to be nonhomogeneous.

General

In many problems involving thermoelastic plates, it is seen that the boundary conditions are nonhomogeneous. Why is this important? In solving linear partial differential equations, separation of variables cannot be used with nonhomogeneous boundary conditions. Fortunately, methods are available to transform either homogeneous or nonhomogeneous partial differential equations with nonhomogeneous boundary conditions to nonhomogeneous partial differential equations with homogeneous boundary conditions, so that separation of variables may be used. A generalized method is presented in the next section.

5.4. General Treatment of Plate Nonhomogeneous Boundary Conditions

Consider a plate with the $y = 0$, b edges simply supported. The governing equation for the lateral deflection is given by Equation (5.14). From Equations (5.21) and (5.22), the boundary conditions are:

$$w(x, 0) = w(x, b) = 0 \tag{5.26}$$

$$\frac{\partial^2 w}{\partial y^2}(x, 0) = -\frac{M^*(x, 0)}{D(1 - v)} = -\frac{M_1^*(x)}{D(1 - v)} \tag{5.27}$$

$$\frac{\partial^2 w}{\partial y^2}(x, b) = -\frac{M^*(x, b)}{D(1 - v)} = -\frac{M_2^*(x)}{D(1 - v)} \tag{5.28}$$

where $M_1^*(x) = M^*(x, 0)$ and $M_2^* = M^*(x, b)$.

We now introduce a function $\Psi(x, y)$, which satisfies homogeneous boundary conditions on the $y = 0$ and $y = b$ edges. Let

$$w(x, y) = \Psi(x, y) + f_1(y)M_1^*(x) + f_2(y)M_2^*(x) . \tag{5.29}$$

where for this problem we take $\Psi(x, y)$ to be of the Levy form:

$$\Psi(x, y) = \sum_{n=1}^{\infty} \phi_n(x) \sin \frac{n\pi y}{b} . \tag{5.30}$$

Also in Equation (5.29) $f_1(y)$ and $f_2(y)$ are to be determined to satisfy the boundary conditions (5.26) through (5.28).

Substituting Equation (5.29) into Equations (5.26) through (5.28) results in

$$w(x, 0) = \Psi(x, 0) + f_1(0)M_1^*(x) + f_2(0)M_2^*(x) = 0$$

$$w(x, b) = \Psi(x, b) + f_1(b)M_1^*(x) + f_2(b)M_2^*(x) = 0$$

$$\frac{\partial^2 w(x, 0)}{\partial y^2} = \frac{\partial^2 \Psi(x, 0)}{\partial y^2} + f_1''(0)M_1^*(x) + f_2''(0)M_2^*(x) = -\frac{M_1^*(x)}{D(1 - v)}$$

$$\frac{\partial^2 w(x, b)}{\partial y^2} = \frac{\partial^2 \Psi(x, b)}{\partial y^2} + f_1''(b)M_1^*(x) + f_2''(b)M_2^*(x) = -\frac{M_2^*(x)}{D(1 - v)} .$$

Since it is required that $\Psi(x, y)$ satisfy homogeneous boundary condition at $y = 0$ and $y = b$, then

$$\Psi(x, 0) = \Psi(x, b) = \frac{\partial^2 \Psi(x, 0)}{\partial y^2} = \frac{\partial^2 \Psi(x, b)}{\partial y^2} = 0 . \tag{5.31}$$

Hence, from the above:

$$\begin{aligned}
f_1(0) &= 0 & f_2(0) &= 0 \\
f_1(b) &= 0 & f_2(b) &= 0 \\
f_1''(0) &= -1/D(1 - v) & f_2''(0) &= 0 \\
f_1''(b) &= 0 & f_2''(b) &= -1/D(1 - v) .
\end{aligned} \tag{5.32}$$

These are the only requirements on $f_1(y)$ and $f_2(y)$. Since there are four conditions on each function each can be assumed to be a third order polynomial.

Let

$$f_1(y) = C_0 + C_1 y + C_2 y^2 + C_3 y^3 \tag{5.33}$$

and

$$f_2(y) = k_0 + k_1 y + k_2 y^2 + k_3 y^3 . \tag{5.34}$$

Substituting Equations (5.33) and (5.34) into (5.32), the result is

$$f_1(y) = \frac{1}{6bD(1 - v)} \ (2b^2 y - 3by^2 + y^3) \tag{5.35}$$

$$f_2(y) = \frac{1}{6bD(1 - v)} \ (b^2 y - y^3) . \tag{5.36}$$

Using Equations (5.35) and (5.36), the substitution of Equation (5.29) into (5.14) results in the following:

$$DV^4 \Psi = p(x, y) - \frac{1}{(1 - v)} \ V^2 M^*$$

$$- V^4 \left\{ \frac{1}{6(1 - v)b} \ (2b^2 y - 3by^2 + y^3) M_1^*(x) \right\}$$

$$- V^4 \left\{ \frac{b^2 y - y^3}{6(1 - v)b} \ M_2^*(x) \right\} . \tag{5.37}$$

Looking at Equation (5.37) it is seen that the original problem, which was Equation (5.14), with nonhomogeneous boundary conditions, Equations (5.26) through (5.28), has been transformed into a problem involving a 'lateral deflection' Ψ, with homogeneous boundary conditions (5.31) and an 'altered loading', given by the right-hand side of Equation (5.37), which we shall now simply write as $H(x, y)$. Hence,

$$DV^4 \Psi = H(x, y) . \tag{5.38}$$

Here $\Psi(x, y)$ is given by Equation (5.30) and $H(x, y)$ must be expanded correspondingly into a Fourier series as

$$H(x, y) = \sum_{n=1}^{\infty} h_n(x) \sin \lambda_n y \quad \text{where} \quad \lambda_n = n\pi/b . \tag{5.39}$$

Substituting (5.30) and (5.39) into Equation (5.38) gives

$$D \sum_{n=1}^{\infty} \{ \phi_n^{iv} - 2\lambda_n^2 \phi_n'' + \lambda_n^4 \phi_n \} \sin \lambda_n y = \sum_{n=1}^{\infty} h_n(x) \sin \lambda_n y .$$

Hence,

$$\phi_n^{iv} - 2\lambda_n^2 \phi_n'' + \phi_n \lambda_n^4 = \frac{h_n(x)}{D} . \tag{5.40}$$

It is seen that this has the same form of the ordinary differential equation in the Chapter 4 discussion of the Levy method. Now the boundary conditions at $x = 0$ and $x = a$ can be considered. For the sake of a specific example, consider them to be simple supported also. Then,

$$w(0, y) = w(a, y) = 0$$

$$M_x(0, y) = M_x(a, y) = 0$$

or

$$\frac{\partial^2 w(0, y)}{\partial x^2} = -\frac{M^*(0, y)}{D(1 - v)}$$

$$\frac{\partial^2 w(a, y)}{\partial x^2} = -\frac{M^*(a, y)}{D(1 - v)} .$$

(5.41)

Substituting Equation (5.29) into the above results in

$$w(0, y) = \Psi(0, y) + f_1(y)M_1^*(0) + f_2(y)M_2^*(0) = 0$$

$$w(a, y) = \Psi(a, y) + f_1(y)M_1^*(a) + f_2(y)M_2^*(a) = 0$$

$$\frac{\partial^2 w(0, y)}{\partial x^2} = \frac{\partial^2 \Psi}{\partial x^2}(0, y) + f_1(y)M_1^{*''}(0) + f_2(y)M_2^{*''}(0) = -\frac{M^*(0, y)}{D(1 - v)}$$

$$\frac{\partial^2 w(a, y)}{\partial x^2} = \frac{\partial^2 \Psi}{\partial x^2}(a, y) + f_1(y)M_1^{*''}(a) + f_2(y)M_2^{*''}(a) = -\frac{M^*(a, y)}{D(1 - v)} .$$

Rearranging the above produces

$$\Psi(0, y) = -[f_1(y)M_1^*(0) + f_2(y)M_2^*(0)]$$

$$\Psi(a, y) = -[f_1(y)M_1^*(a) + f_2(y)M_2^*(a)]$$

(5.42)

$$\frac{\partial^2 \Psi(0, y)}{\partial x^2} = -\left[\frac{M^*(0, y)}{D(1 - v)} + f_1(y)M_1^{*''}(0) + f_2(y)M_2^{*''}(0)\right]$$

$$\frac{\partial^2 \Psi(a, y)}{\partial x^2} = -\left[\frac{M^*(a, y)}{D(1 - v)} + f_1(y)M_1^{*''}(a) + f_2(y)M_2^{*''}(a)\right] .$$

Remembering that $\Psi(x, y)$ is given by equation (5.30), it is logical to make the following expansions:

$$f_1(y) = \sum_{n=1}^{\infty} A_n \sin \lambda_n y \quad M^*(0, y) = \sum_{n=1}^{\infty} C_n \sin \lambda_n y$$

(5.43)

$$f_2(y) = \sum_{n=1}^{\infty} B_n \sin \lambda_n y \quad M^*(a, y) = \sum_{n=1}^{\infty} E_n \sin \lambda_n y .$$

Here, A_n, B_n, C_n, and E_n are easily found. Substituting Equations (5.43) and (5.30) into (5.42) and equating all coefficients results in

$$\phi_n(0) = -[A_n M_1^*(0) + B_n M_2^*(0)]$$

$$\phi_n(a) = -[A_n M_1^*(a) + B_n M_2^*(a)]$$

$$\phi_n''(0) = -\left[\frac{C_n}{D(1-v)} + A_n M_1^{*''}(0) + B_n M_2^{*''}(0)\right]$$

$$\phi_n''(a) = -\left[\frac{E_n}{D(1-v)} + A_n M_1^{*''}(a) + B_n M_2^{*''}(a)\right].$$

(5.44)

Hence, these four boundary values provide the necessary information to determine the constants in the solution of Equation (5.40), which is

$$\phi_n(x) = (K_1 + K_2 x)\cosh\lambda_n x + (K_3 + K_4 x)\sinh\lambda_n x + \eta_n(x).$$ (5.45)

Here, $\eta_n(x)$ is the particular solution. Using this then, Equation (5.30) is completely solved, and in turn Equation (5.29) is solved. Subsequently, stress couples, shear resultants, and stresses can be determined everywhere using Equations (5.19) and (5.20).

This general approach can be used to solve any plate problem involving nonhomogeneous boundary conditions.

5.5. Thermoelastic Effects on Beams

For the thermoelastic beam, it can be easily shown from Chapters 2 and 3, and Section 5.2 that the governing differential equations are as follows:

$$P + P^* = EA\,\frac{du_0}{dx}$$ (5.46)

$$M_b + M_b^* = -EI\,\frac{d^2 w}{dx^2}$$ (5.47)

$$V_b = -EI\,\frac{d^3 w}{dx^3} - \frac{dM^*}{dx}$$ (5.48)

$$EI\,\frac{d^4 w}{dx^4} = q(x) - \frac{d^2 M^*}{dx^2}$$ (5.49)

$$\sigma_x = \frac{1}{A}\,[P + P^*] + \frac{z}{I}\,[M_b + M^*] - E\alpha\Delta T$$ (5.50)

$$P^* = \int_{-h/2}^{+h/2} Eb\alpha\Delta T\,dz, \quad M^* = \int_{-h/2}^{+h/2} Eb\alpha\Delta Tz\,dz.$$ (5.51)

Using these equations, solutions are easily obtained, analogous to those obtained in Chapter 3. As for plates, the expressions for boundary conditions of simple supported and free edges will be nonhomogeneous, but in this case (unlike plates) this causes no particular problem, because ordinary differential equations are involved, not partial differential equations, hence, separation of variables is not needed.

5.6. Self-Equilibration of Thermal Stresses

In Section 5.1 the two mechanisms by which thermal stresses are introduced into a thermoelastic solid body are discussed, namely, by displacement restrictions caused by the boundary conditions, the other by the introduction of the isothermal strains in Equation 5.2 in order that $\varepsilon_{ij_{total}}$ satisfy compatibility when the thermal strains, $\varepsilon_{ij_{th}}$ do not satisfy compatibility.

One other physical phenomenon occurs that is very important in the structural mechanics of planar bodies such as beams and plates, namely, if the planar body is not restricted by the boundary conditions the thermals stresses in the body are self-equilibrating: i.e., the average stress across the thickness is zero,

$$\sigma_{i_{avg}} = \frac{1}{h} \int_{-h/2}^{+h/2} \sigma_i \, dz = 0 \quad (i = x, y) \tag{5.52}$$

This can be exemplified in the easiest and shortest way by considering a load free beam lying on a friction free surface and heated or cooled such that at any time t, the temperature change is given by:

$$\Delta T(z) = \sum_{n=0}^{\infty} a_n \left(\frac{z}{h}\right)^n = a_0 + a_1 \frac{z}{h} + a_2 \frac{z^2}{h^2} + \dots \tag{5.53}$$

It is seen that the term a_0 is merely a uniform heating or cooling of the entire beam; $(a_0 + a_1 z)$ is a steady state heat transfer situation; and the entire expression represents a temperature situation that involves additional terms occurs during transient heating or cooling and/or internal heat generation. In the following example, it is sufficient to consider only the first three terms of (5.53) to illustrate the point. From (5.53) and (5.51),

$$P^* = EA\alpha \left[a_0 + \frac{a_2}{12} \right] \quad M^* = \frac{E\alpha bh^2}{12} a_1 . \tag{5.54}$$

For the illustrative problem, the lateral deflection is taken as

$$w(x) = C_0 + C_1 x + C_2 x^2 + C_3 x^3 \tag{5.55}$$

where this form is the solution to Equation (5.49) for this case. The boundary conditions for this example are seen to be simply supported: i.e.,

$$w(0, L) = M_b(0, L) = 0 . \tag{5.56}$$

With these boundary conditions, the constants of (5.55) are found to be

$$C_0 = 0 \quad C_1 = \frac{\alpha a_1 L}{2A} \quad C_2 = -\frac{\alpha a_1}{2A} \quad C_3 = 0 \tag{5.57}$$

and the lateral deflection is seen to be

$$w(x) = \frac{\alpha a_1}{2A} \, x(L - x).$$ (5.58)

It is seen that only if a_1 (or any a_n, n odd) is non-zero will the beam deflect at all. Also from (5.50), it is seen that

$$\sigma_x = E\alpha \, \frac{a_2}{12} \left[1 - \frac{12z^2}{h^2} \right] \quad (i = x, y).$$ (5.59)

It is extremely important to see that stresses will occur in the beam if and only if a_2 (or, a_n, $n \geqslant 2$) is non-zero. Therefore, it is seen that for a steady state temperature distribution $(a_n, n \geqslant 2 = 0)$ the beam is stress free, whether a deflection occurs or not. Substituting (5.59) into (5.52) results in, for all cases,

$$\sigma_{i_{avg}} = 0.$$ (5.60)

Therefore, in those cases of no boundary conditions constraining the thin beam or plate, the average stress across the thickness is zero, i.e., the stresses are self-equilibrating.

5.7. References

5.1. Boley, B. A. and J. H. Wiener, *Theory of Thermal Stresses*, New York: John Wiley and Sons, Inc., 1962.
5.2. Johns, D. J., *Thermal Stress Analyses*, Pergamon Press, 1965.
5.3. Nowinski, J. L., *Theory of Thermoelasticity with Applications*, Alphen aan den Rijn; Sijthoff and Noordhoff Publishers, 1978.
5.4. Vinson, J. R. and R. L. Sierakowski, *The Behavior of Structures Composed of Composite Materials*, Dordrecht: Martinus-Nijhoff Publishers, 1986.

5.8. Problems

5.1. A flat structural panel on the wing of a supersonic fighter is considered to be unstressed at 70 °F. After a considerable time at cruise speed such that a steady state temperature distribution is reached, the temperature on the heated side is measured to be 140 °F, the temperature on the cooler side is measured at 80 °F, and the temperature gradient through the plate is considered to be linear. Calculate the thermal stress resultant, N^*, and the thermal stress couple, M^*, where for aluminum $E = 10 \times 10^6$ psi, $\alpha = 10 \times 10^{-6}$ in/in/°F, and $\nu = 0.3$.

5.2. The same aluminum panel as in Problem 1 is now heated symmetrically from both the top side and the bottom side. After 10 seconds thermocouples placed on both surfaces of the panel read 160 °F, and a thermocouple at the mid-surface reads 180 °F. Assuming the temperature profile in the panel to be parabolic (i.e., a second degree polynomial), what is the thermal stress resultant N^* and the thermal stress couple M^* at this time?

5.3. In Problem 5.1 at what location across the plate thickness are σ_x and σ_y a maximum and what is that stress assuming $N_x = N_y = M_x = M_y = 0$?

5.4. In Problem 5.2 at what location across the plate thickness are σ_x and σ_y a maximum value and what is the value; assuming $N_x = N_y = M_x = M_y = 0$?

5.5. An aluminum panel 0.4″ thick, and stress free at 70 °F is subjected to transient heating on one of its surface such that $T = T(z)$ only as in the previous problems. Thermocouples record at a critical time that $T(h/2) = 170$ °F, $T(0) = 130$ °F, and $T(-h/2) = 100$ °F. Assuming a polynomial temperature distribution calculated N^*, M^* and $\sigma_{x_{max}} = \sigma_{y_{max}}$ assuming N_x, N_y, M_x, and $M_y = 0$?

5.6. Thermocouples are used to measure the temperature profile through a two inch thick plate, through three measurements: one on the upper surface, one at the mid-surface, and one on the lower surface. At one specific time the measurements were:

Location	Actual Temperature Measured
$h = +1''$	200 °F
$h = 0''$	110 °F
$h = -1''$	80 °F

If the stress free temperature is 60 °F, calculate N^* and M^*.

5.7. A plate is heated from both the top and bottom such that at a certain time, three thermocouples read $T(+h/2) = 300$ °F, $T(0) = 80$ °F, and $T(-h/2) = 300$ °F. If the stress free temperature is 70 °F, calculate N^* and M^* for a plate that is 2″ in thickness. This aluminum plate has $E = 10 \times 10^6$ psi and $\alpha = 10 \times 10^{-6}$ in/in/°F.

5.8. A thin walled structure $\frac{1}{2}''$ thick, i.e. $-\frac{1}{4}'' \leqslant z \leqslant +\frac{1}{4}''$, is composed of an aluminium with properties $E = 10 \times 10^6$ psi and $\alpha = 10 \times 10^{-6}$ in/in/°F. As a certain time t, thermocouples on the top, at midsurface and at the bottom record 90 °F, 100 °F and 150 °F. If the stress free temperature is 70 °F, determine the equation for ΔT to perform w subsequent thermoelastic analysis.

Circular Plates

6.1. Introduction

In previous chapters, attention has been focused on rectangular plates. However, circular plate structural elements are encountered in all phases of engineering. It is, therefore, necessary to develop an understanding of the behavior of circular plates.

Consider the following element from a circular plate (Figure 6.1), with positive directions of stresses and deflections as shown.

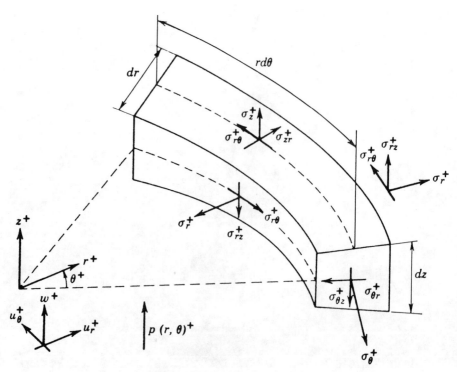

Figure 6.1. Circular plate element.

6.2. Derivation of the Governing Equations

The equations of elasticity can be derived in a circular cylindrical coordinate system, or could be obtained by transforming the elasticity equations given in Chapter 1 through the use of the relationships:

$$x = r \cos \theta, \quad y = r \sin \theta \quad \text{and} \quad x^2 + y^2 = r^2 .$$

However, they are merely presented here in their final form.

Equilibrium Equations in Circular Cylindrical Coordinates

$$\frac{\partial \sigma_r}{\partial r} + \frac{1}{r} \frac{\partial \sigma_{r\theta}}{\partial \theta} + \frac{\partial \sigma_{rz}}{\partial z} + \frac{\sigma_r - \sigma_\theta}{r} = 0 \tag{6.1}$$

$$\frac{\partial \sigma_{r\theta}}{\partial r} + \frac{1}{r} \frac{\partial \sigma_\theta}{\partial \theta} + \frac{\partial \sigma_{\theta z}}{\partial z} + \frac{z}{r} \sigma_{r\theta} = 0 \tag{6.2}$$

$$\frac{\partial \sigma_{rz}}{\partial r} + \frac{1}{r} \frac{\partial \sigma_{\theta z}}{\partial \theta} + \frac{\partial \sigma_z}{\partial z} + \frac{1}{r} \sigma_{rz} = 0 . \tag{6.3}$$

Stress-Strain Relations (after using plate assumptions)

$$\varepsilon_r = \frac{1}{E} [\sigma_r - v\sigma_\theta] \tag{6.4}$$

$$\varepsilon_\theta = \frac{1}{E} [\sigma_\theta - v\sigma_r] \tag{6.5}$$

$$\varepsilon_{r\theta} = \frac{1}{2G} \sigma_{r\theta} \tag{6.6}$$

$$\sigma_z = \varepsilon_{rz} = \varepsilon_{\theta z} = 0 . \tag{6.7}$$

Strain-Displacement Relations, General

$$\varepsilon_r = \frac{\partial u_r}{\partial r} \; ; \quad \varepsilon_\theta = \frac{1}{r} \frac{\partial u_\theta}{\partial \theta} + \frac{u_r}{r} \; ; \quad \varepsilon_z = \frac{\partial w}{\partial z} \tag{6.8}$$

$$\varepsilon_{r\theta} = \frac{1}{2} \left(\frac{1}{r} \frac{\partial u_r}{\partial \theta} + \frac{\partial u_\theta}{\partial r} - \frac{u_\theta}{r} \right) \tag{6.9}$$

$$\varepsilon_{\theta z} = \frac{1}{2} \left(\frac{\partial u_\theta}{\partial z} + \frac{1}{r} \frac{\partial w}{\partial \theta} \right) \tag{6.10}$$

$$\varepsilon_{rz} = \frac{1}{2} \left(\frac{\partial w}{\partial r} + \frac{\partial u_r}{\partial z} \right) . \tag{6.11}$$

Of course, for a classical plate $\varepsilon_z = \varepsilon_{rz} = \varepsilon_{\theta z} = 0$ in (6.8), (6.10), and (6.11) above.

Similar to the case of rectangular plates, stress resultants, stress couples and shear resultants are defined as follows:

$$\begin{Bmatrix} N_r \\ N_\theta \\ N_{r\theta} \end{Bmatrix} = \int_{-h/2}^{+h/2} \begin{Bmatrix} \sigma_r \\ \sigma_\theta \\ \sigma_{r\theta} \end{Bmatrix} dz \tag{6.12}$$

$$\begin{Bmatrix} M_r \\ M_\theta \\ M_{r\theta} \end{Bmatrix} = \int_{-h/2}^{+h/2} \begin{Bmatrix} \sigma_r \\ \sigma_\theta \\ \sigma_{r\theta} \end{Bmatrix} z\, dz \tag{6.13}$$

$$\begin{Bmatrix} Q_r \\ Q_\theta \end{Bmatrix} = \int_{-h/2}^{+h/2} \begin{Bmatrix} \sigma_{rz} \\ \sigma_{\theta z} \end{Bmatrix} dz \,. \tag{6.14}$$

In developing the governing equations for a circular plate, one proceeds as in Chapter 2, multiplying Equations (6.1) through (6.3) by dz and integrating the equations across the thickness of the plate: then multiplying (6.1) and (6.2) by $z\, dz$ and again integrating these across the plate thickness. The results are:

$$\frac{\partial N_r}{\partial r} + \frac{1}{r}\frac{\partial N_{r\theta}}{\partial \theta} + \frac{N_r - N_\theta}{r} = 0 \tag{6.15}$$

$$\frac{\partial N_{r\theta}}{\partial t} + \frac{1}{r}\frac{\partial N_\theta}{\partial \theta} + \frac{2}{r} N_{r\theta} = 0 \tag{6.16}$$

$$\frac{\partial Q_r}{\partial r} + \frac{1}{r}\frac{\partial Q_\theta}{\partial \theta} + \frac{1}{r} Q_r + p(r, \theta) = 0 \tag{6.17}$$

$$\frac{\partial M_r}{\partial r} + \frac{1}{r}\frac{\partial M_{r\theta}}{\partial \theta} + \frac{M_r - M_\theta}{r} - Q_r = 0 \tag{6.18}$$

$$\frac{\partial M_{r\theta}}{\partial r} + \frac{1}{r}\frac{\partial M_\theta}{\partial \theta} + \frac{2}{r} M_{r\theta} - Q_\theta = 0 \,. \tag{6.19}$$

Similar to Chapter 2, for the bending of a circular plate, displacements are taken in the form of

$$u_r = u_{0r} + \bar{\alpha}z, \quad u_\theta = u_{0\theta} + \beta z, \quad \text{and} \quad w = w(r, \theta) \,. \tag{6.20}$$

Since in a classical circular plate $\varepsilon_{rz} = \varepsilon_{\theta z} = 0$, substituting (6.20) into

Equations (6.10) and (6.11) results in

$$\bar{\alpha} = -\frac{\partial w}{\partial r} \quad \text{and} \quad \beta = -\frac{1}{r}\frac{\partial w}{\partial \theta} \quad \text{or}$$

$$u_r = u_{0r} - z\frac{\partial w}{\partial r}, \quad u_\theta = u_{0\theta} - \frac{z}{r}\frac{\partial w}{\partial \theta}. \tag{6.21}$$

From Equations (6.21), (6.8), and (6.4),

$$\varepsilon_r = \frac{\partial u_r}{\partial r} = \frac{1}{E}[\sigma_r - v\sigma_\theta] = \frac{\partial u_{0r}}{\partial t} - z\frac{\partial^2 w}{\partial r^2}. \tag{6.22}$$

From Equations (6.21), (6.8), and (6.5),

$$\varepsilon_\theta = \frac{1}{r}\frac{\partial u_\theta}{\partial \theta} + \frac{u_r}{r} = \frac{1}{E}[\sigma_\theta - v\sigma_r] =$$

$$= \frac{1}{r}\frac{\partial u_{0\theta}}{\partial \theta} + \frac{u_{0r}}{r} - \frac{z}{r^2}\frac{\partial^2 w}{\partial \theta^2} - \frac{z}{r}\frac{\partial w}{\partial r}. \tag{6.23}$$

From Equations (6.21), (6.9), and (6.6),

$$\varepsilon_{r\theta} = \frac{1}{2G}\sigma_{r\theta} = \frac{1+v}{E}\sigma_{r\theta} =$$

$$= \frac{1}{2}\left(\frac{1}{r}\frac{\partial u_{0r}}{\partial \theta} - \frac{\partial u_{0\theta}}{\partial r} - \frac{u_{0\theta}}{r}\right) - \frac{2z}{r}\frac{\partial^2 w}{\partial r\,\partial \theta} + \frac{2z}{r^2}\frac{\partial w}{\partial \theta}. \tag{6.24}$$

Multiplying Equations (6.22) through (6.24) by dz, and integrating across the thickness of the plate, then multiplying them by $z\,dz$ and again integrating them across the plate thickness and with some algebraic manipulation the stress resultant in-plane displacement relations and moment-curvature relations for a circular plate evolve (for the case of no surface shear stresses).

$$N_r = K\left[\frac{\partial u_{0r}}{\partial r} + \frac{v}{r}\frac{\partial u_{0\theta}}{\partial \theta} + \frac{vu_{0r}}{r}\right] \tag{6.25}$$

$$N_\theta = K\left[\frac{1}{r}\frac{\partial u_{0\theta}}{\partial \theta} + \frac{u_{0r}}{r} + v\frac{\partial u_{0r}}{\partial r}\right] \tag{6.26}$$

$$N_{r\theta} = K(1-v)\left[\frac{1}{r}\frac{\partial u_{0r}}{\partial \theta} + \frac{\partial u_{0\theta}}{\partial r} - \frac{u_{0\theta}}{r}\right] \tag{6.27}$$

$$M_r = -D\left[\frac{\partial^2 w}{\partial r^2} + \frac{v}{r}\frac{\partial w}{\partial r} + \frac{v}{r^2}\frac{\partial^2 w}{\partial \theta^2}\right] \tag{6.28}$$

$$M_\theta = -D\left[\frac{1}{r^2}\frac{\partial^2 w}{\partial \theta^2} + \frac{1}{r}\frac{\partial w}{\partial r} + v\frac{\partial^2 w}{\partial r^2}\right] \tag{6.29}$$

$$M_{r\theta} = -D(1-v)\left[\frac{1}{r}\frac{\partial^2 w}{\partial r\,\partial\theta} - \frac{1}{r^2}\frac{\partial w}{\partial\theta}\right] \tag{6.30}$$

where again $K = Eh/(1 - v^2)$ and $D = Eh^3/12(1 - v^2)$, the in-plane stiffness and flexural stiffness, respectively.

Solving (6.18) and (6.19) for Q_r and Q_θ, and substituting the result into Equation (6.17) provides an equation involving $M_r, M_\theta, M_{r\theta}$ and $p(r, \theta)$. Substituting Equations (6.28) through (6.30) into that equation results in the final governing differential equation for the bending of a circular plate, again the biharmonic equation.

$$D\nabla^4 w = p(r, \theta) \tag{6.31}$$

where

$$\nabla^2(\ ) = \frac{\partial^2(\ )}{\partial r^2} + \frac{1}{r}\frac{\partial(\ )}{\partial r} + \frac{1}{r^2}\frac{\partial^2(\ )}{\partial\theta^2}\ . \tag{6.32}$$

Similarly, substituting (6.25) through (6.27) into (6.15) and (6.16) produces the equations for the stretching of a circular plate.

$$\nabla^4 u_{0r} = 0$$
$$\nabla^4 u_{0\theta} = 0\ . \tag{6.33}$$

Of course once the plate solution is obtained, the stresses within the plate are given by:

$$\sigma_r = \frac{N_r}{h} + \frac{M_r z}{h^3/12}$$

$$\sigma_\theta = \frac{N_\theta}{h} + \frac{M_\theta z}{h^3/12}$$

$$\sigma_{r\theta} = \frac{N_{r\theta}}{h} + \frac{M_{r\theta} z}{h^3/12} \tag{6.34}$$

$$\sigma_{rz} = \frac{3Q_r}{2h}\left[1 - \left(\frac{z}{h/2}\right)^2\right]$$

$$\sigma_{\theta z} = \frac{3Q_\theta}{2h}\left[1 - \left(\frac{z}{h/2}\right)^2\right]\ .$$

For the case of surface shear stresses, the last two expressions above would be modified by the analogous expressions of Chapter 2, simply modified by changing x and y subscripts to r and θ.

Furthermore, to consider a thermoelastic circular plate, one merely adds append-ages to Equations (6.25), (6.26), (6.28), (6.29), (6.31), (6.33), and the first two of (6.34), identical to the last terms of the N_x, N_y, M_x, M_y expressions of Chapter 5, and the modifications for σ_x and σ_y with obvious subscripts changes.

In the general case of axial asymmetry, the solution of Equation (6.31) results in Bessel functions and modified Bessel functions of the first and second kinds. Such problems will not be treated herein, but are treated in depth in various other texts dealing with circular plates.

6.3. Axially Symmetric Circular Plates

When the plate is continuous in the θ direction, (i.e., is in the region $0 \leqslant \theta \leqslant 2\pi$), when the loading is not a function of θ, and when the boundary conditions do not vary around the circumference, the plate problem is said to be axially symmetric, and the following simplifications can be made:

$$\frac{\partial(\ )}{\partial\theta} = \frac{\partial^2(\ )}{\partial\theta^2} = M_{r\theta} = Q_\theta = 0 . \tag{6.35}$$

The previous equations for the bending of a circular plate can therefore be simplified to the following, where primes denote differentiation with respect to r.

$$M_r = -D\left(w'' + \frac{v}{r}w'\right) \tag{6.36}$$

$$M_\theta = -D\left(\frac{1}{r}w' + vw''\right) \tag{6.37}$$

$$Q_r = -D(\nabla^2 w)' \tag{6.38}$$

$$D\nabla^4 w = p(r) \tag{6.39}$$

where

$$\nabla^2(\ ) = (\ )'' + \frac{1}{r}(\ )' = \frac{1}{r}\frac{d}{dr}\left[r\frac{d(\ )}{dr}\right]. \tag{6.40}$$

Interestingly, Equation (6.39) can therefore be written as,

$$\nabla^4 w = \frac{1}{r}\frac{d}{dr}\left\{r\frac{d}{dr}\left[\frac{1}{r}\frac{d}{dr}\left(r\frac{dw}{dr}\right)\right]\right\} = \frac{p(r)}{D} . \tag{6.41}$$

6.4. Solutions for Axially Symmetric Circular Plates

Equation (6.41) can be made dimensionless by normalizing both the radial coor-dinate, r, and the lateral deflection, w, with respect to the radius of the circular plate, *a*, as follows:

$$\bar{r} = r/a, \quad \bar{w} = w/a.$$

(6.42)

Using (6.42) above, Equation (6.41) can be written as

$$\frac{1}{\bar{r}}\frac{d}{d\bar{r}}\left\{\bar{r}\frac{d}{dr}\left[\frac{1}{\bar{r}}\frac{d}{d\bar{r}}\left(\bar{r}\frac{d\bar{w}}{d\bar{r}}\right)\right]\right\} = \frac{p(\bar{r})a^3}{D}.$$

(6.43)

One can proceed to obtain the homogeneous solution of Equation (6.43) above, by setting the right hand side equal to zero, and proceeding to integrate the left hand side, where below, C_0, C_1, C_2, and C_3 are the resulting constants of integration used to satisfy boundary conditions.

Multiplying the homogeneous portion of Equation (6.43) by $\bar{r}$, then integrating once yields

$$\frac{d}{d\bar{r}}\left[\frac{1}{\bar{r}}\frac{d}{d\bar{r}}\left(\bar{r}\frac{d\bar{w}}{d\bar{r}}\right)\right] = \frac{C_0}{\bar{r}}.$$

Integrating once and multiplying both sides by *r* provides

$$\frac{d}{d\bar{r}}\left(\bar{r}\frac{d\bar{w}}{d\bar{r}}\right) = C_0\bar{r}\ln\bar{r} + C_1\bar{r}.$$

To integrate the first term in the right hand side, let $\ln\bar{r} = y$, hence, $\bar{r} = e^y$, and $d\bar{r} = e^y\,dy$. Therefore,

$$\int \bar{r}\ln\bar{r}\,d\bar{r} = \int ye^{2y}\,dy = \frac{ye^{2y}}{2} - \frac{e^{2y}}{4}.$$

Therefore, integrating the expression above, and dividing the results by $\bar{r}$ gives

$$\frac{d\bar{w}}{d\bar{r}} = C_0\left[\frac{\bar{r}\ln\bar{r}}{2} - \frac{\bar{r}}{4}\right] + \frac{C_1\bar{r}}{2} + \frac{C_2}{\bar{r}}$$

and finally, one more integration produces

$$\bar{w} = \frac{C_0}{2}\left[\frac{\bar{r}^2\ln\bar{r}}{2} - \frac{\bar{r}^2}{4}\right] - \frac{C_0\bar{r}^2}{8} + \frac{C_1\bar{r}^2}{4} + C_2\ln\bar{r} + C_3.$$

This final form of the homogeneous solution can be written more succinctly as

$$\bar{w} = A + B\ln\bar{r} + C\bar{r}^2 + E\bar{r}^2\ln\bar{r}.$$

(6.44)

Returning to Equation (6.43), the particular solution can be written as:

$$\bar{w}_p = \int\frac{1}{\bar{r}}\int\bar{r}\int\frac{1}{\bar{r}}\int\frac{p(\bar{r})a^3}{D}\bar{r}\,d\bar{r}\,d\bar{r}\,d\bar{r}\,d\bar{r}.$$

(6.45)

Thus, the total solution for any circular plate under axially symmetric loading is given by Equations (6.44) and (6.45). It is easy to show that the particular solution for a plate with a uniform lateral load is:

$$\overline{w}_p = \frac{p_0 a^3 \overline{r}^3}{64D} \quad \text{or} \quad w_p = \frac{p_0 r^4}{64D} . \tag{6.46}$$

For ease of calculation, the following quantities are given explicitly for the circular plate of radius a with uniform lateral loading $p(r) = p_0$.

$$\overline{w} = A + B \ln \overline{r} + C\overline{r}^2 + E\overline{r}^2 \ln \overline{r} + \frac{p_0 a^3 \overline{r}^4}{64D} \tag{6.47}$$

$$\frac{d\overline{w}}{d\overline{r}} = \frac{B}{\overline{r}} + 2C\overline{r} + E[2\overline{r} \ln \overline{r} + \overline{r}] + \frac{p_0 a^3 \overline{r}^3}{16D} \tag{6.48}$$

$$M_r = -\frac{D}{a}\left[-\frac{B}{\overline{r}^2}(1 - v) + 2C(1 + v) + 2E(1 + v)\ln \overline{r} \right.$$

$$\left. + (3 + v)E \right] - \frac{p_0 a^2 \overline{r}^2 (3 + v)}{16} \tag{6.49}$$

$$Q_r = -\frac{D}{a^2}\left[\frac{4E}{\overline{r}} \right] - \frac{p_0 a \overline{r}}{2} \tag{6.50}$$

$$M_\theta = -\frac{D}{a}\left[\frac{B(1 - v)}{\overline{r}^2} + 2C(1 + v) + 2E \ln \overline{r}(1 + v) + E(1 + 3v) \right]$$

$$- \frac{p_0 a^2 \overline{r}^2 (1 + 3v)}{16} . \tag{6.51}$$

For other lateral loadings, the last terms only in each expression above would be changed, the homogeneous solution remains the same.

6.5. Circular Plate, Simply Supported at the Outer Edge, Subjected to a Uniform Lateral Loading, p_0

For the plate which is continuous from $0 \leqslant r \leqslant a$, and which contains no concentrated loading at $r = 0$, it is easy to see that $B = E = 0$; otherwise the lateral deflection and transverse shear resultant would be infinite at $r = 0$. At $\overline{r} = 1$ or $r = a$, the boundary conditions are

$$\overline{w}(1) = 0$$

$$M_r(1) = 0$$

Hence,

$$A = \frac{5 + v}{1 + v}\frac{p_0 a^3}{64D}, \quad C = -\frac{p_0 a^3 (3 + v)}{32D(1 + v)}$$

and

$$w(\bar{r}) = \frac{p_0 a^4}{64D} \left[\frac{(5 + v)}{(1 + v)} - \frac{2(3 + v)}{(1 + v)} \, \bar{r}^2 + \bar{r}^4 \right]. \tag{6.52}$$

6.6. Circular Plate, Clamped at the Outer Edge, Subjected to a Uniform Lateral Loading, p_0

Again $B = E = 0$. At the outer edge, $w(1) = 0$ and $\partial \bar{w}(1)/\partial \bar{r} = 0$. Hence,

$$A = \frac{p_0 a^3}{64D} \quad \text{and} \quad C = -\frac{p_0 a^3}{32D} \, .$$

Thus,

$$w(\bar{r}) = \frac{p_0 a^4}{64D} \, [1 - 2\bar{r}^2 + \bar{r}^4] \, . \tag{6.53}$$

6.7. Annular Plate, Simply Supported at the Outer Edge, Subjected to a Stress Couple, M, at the Inner Boundary

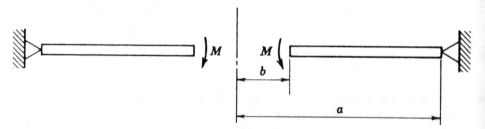

Figure 6.2. Annular circular plate.

Remembering that $r = r/a$, and defining $s = b/a$, the governing differential equation in this case with no lateral load, $p(r)$, is

$$\nabla^4 w = 0$$

and the boundary conditions are:

$$\bar{w}(1) = 0 \quad M_r(s) = M$$

$$M_r(1) = 0 \quad Q_r(s) = 0 \, .$$

The lateral deflection is found to be

$$w(r) = \frac{Ma^2 s^2 \ln \bar{r}}{D(1 - v)(1 - s^2)} - \frac{Ma^2}{2D(1 + v)} \left(\frac{s^2}{1 - s^2} \right) (1 - \bar{r}^2) \, . \tag{6.54}$$

6.8. Annular Plate, Simply Supported at the Outer Edge, Subjected to a Shear Resultant, Q_0, at the Inner Boundary

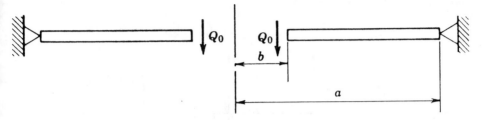

Figure 6.3. Annular circular plate.

Again, the governing differential equation is $\nabla^4 w = 0$, and the boundary conditions are,

$$\overline{w}(1) = 0 \quad M_r(s) = 0$$

$$M_r(1) = 0 \quad Q_r(s) = Q_0$$

The solution is:

$$w(\bar{r}) = \frac{Q_0 s a^3}{4D} \left[-\frac{(3 + v)(1 - \bar{r}^2)}{2(1 + v)} + \frac{s^2 \ln s}{(1 - s^2)}(1 - \bar{r}^2) \right.$$

$$\left. -\frac{2(1 + v)}{(1 - v)}\left(\frac{s^2}{1 - s^2}\right) \ln s \ln \bar{r} - \bar{r}^2 \ln \bar{r} \right]. \tag{6.55}$$

6.9. General Remarks

Of course the results given in Section 6.5 through 6.8 can be superimposed to form the solutions to other problems. Suppose the plate is subjected to a stress couple, M, on the inner boundary as well as a transverse shear resultant, Q_0, acting also at the inner edge, as shown in Figure 6.4.

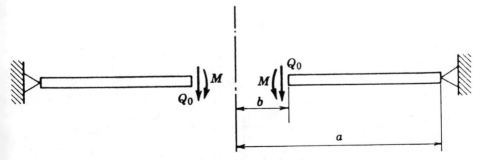

Figure 6.4.. Annular circular plate.

The solution is the sum of (6.54) and (6.55). All other stress quantities are found by substituting this sum into (6.36) through (6.38).

Another example of using the previous examples as building blocks, consider the problem shown in Figure 6.5.

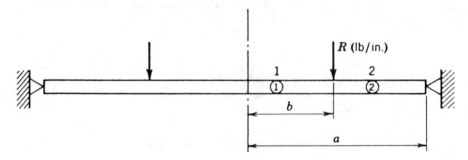

Figure 6.5.. Circular plate with a ring load.

This simply supported circular plate is subjected to a ring load of R (lbs/in. of circumference). To solve this problem one first divides the plate problem into two parts: an inner solution 1 extending over the region $0 \leqslant r \leqslant b$, and an outer solution 2 over the region $b \leqslant r \leqslant a$. In each case the governing equation is

$$\nabla^4 w_1 = 0 \quad \text{and} \quad \nabla^4 w_2 = 0$$

and eight boundary conditions are needed. Since there is no lateral load $p(r)$, and the solution to each equation with suitable subscript, is (6.47) with $p_0 = 0$ (i.e., the homogeneous solution). From the reasoning of Sections 6.5 and 6.6, it is seen that $B_1 = E_1 = 0$. Likewise from the reasoning of Sections 6.5, 6.7, and 6.8, at $r = a$ or $r = 1$, $\overline{w}(1) = M_r(1) = 0$.

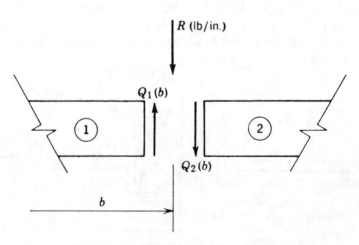

Figure 6.6. Equilibrium of plate with ring load.

At the junction of the two plate segments, it is obvious that the lateral deflection, the slopes, and the stress couples must be equal for both plate segments; hence

$$w_1(b) = w_2(b) \quad \text{or} \quad \overline{w}_1(s) = \overline{w}_2(s)$$

$$\frac{dw_1(b)}{dr} = \frac{dw_2(b)}{dr} \quad \text{or} \quad \frac{d\overline{w}_1(s)}{d\overline{r}} = \frac{d\overline{w}_2(s)}{d\overline{r}}$$

$$M_{r_1}(b) = M_{r_2}(b) \quad \text{or} \quad M_{r_1}(s) = M_{r_1}(s) .$$

For the eighth and last boundary condition is obtained by looking closely at the shear condition at $r = b$, as seen in Figure 6.6.

Hence, $R = Q_2(b) - Q_1(b)$ is the eighth boundary condition. If one has either a discontinuity in load or a discontinuity in plate thickness, one must divide the plate into two segments. Examples of such problems are shown in Figure 6.7.

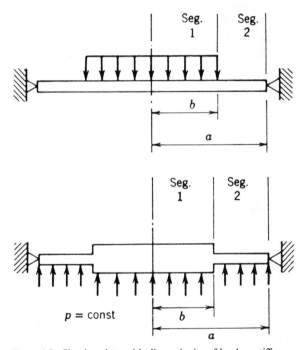

Figure 6.7. Circular plate with discontinuity of load or stiffness.

It should be noted that in the example at the left the lateral load over Segment 1 is a negative number, and the loading in the right example is a positive number, since $p(r)$ is positive in the positive direction of the lateral deflection, w. Further it should be noted that in each of these examples, because there is no concentrated load, R, as in the previous example the eighth boundary condition here is $Q_1(b) = Q_2(b)$.

Of course if one had n structural and/or loading discontinuities, one then must use $(n + 1)$ segments, and $4n$ boundary conditions.

Use of Equations (6.47) through (6.51), with the proper last terms (the particular solution) obtained through solving (6.45) reduces the problems to a straightforward procedure. Subsequently, stresses are found through (6.34).

6.10. Problems

6.1. The circular flooring is a silo of radius a is solidly supported at the walls such that the floor plate is considered to be clamped. If grain is poured onto the floor such that the floor loading is triangular in cross section as shown below what is the expression for the deflection at the center of the floor?

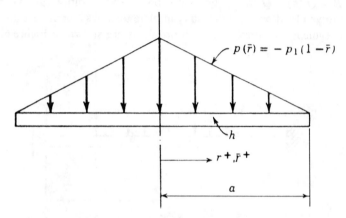

Figure 6.8. Circular plate with varying load.

6.2. Consider a circular plate of radius a, shown on Figure 6.9, loaded by an edge couple $M_r = M$ at the outer edge. Find the value of the stress couples M_r and M throughout the plate.

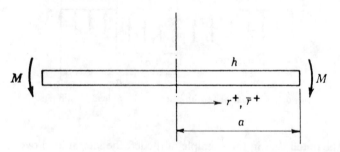

Figure 6.9. Circular plate with edge moment.

6.3. The flat head of a piston in an internal combustion engine is considered to be a plate of radius a, where the center support to the connecting rod is of radius b. If the maximum down pressure is uniform of magnitude $p(r) = -p_0$, determine the location and magnitude of the maximum bending stress? Assume the head is clamped on both edges.

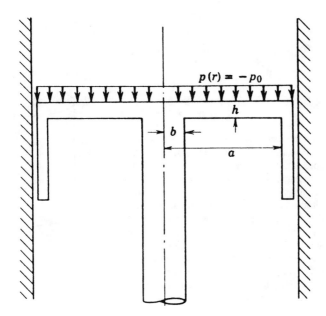

Figure 6.10. Piston in cylinder.

6.4. A certain pressure transducer is made by the pressure in a chamber deflecting a thin circular plate, a rigid member joined to it at the center thus being deflected, and through a linkage mechanism shown, deflecting an arrow on a gage to the right. Determine the expression to relate the deflection of the gage to the pressure p_0 in the chamber. Assume that the rigid piece nor the linkages affect the deflection of the plate by their presence. Assume the circular plate is simply supported.

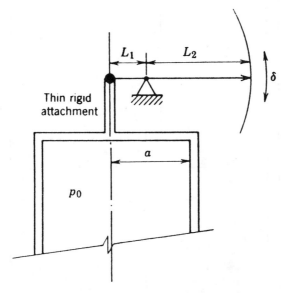

Figure 6.11. Pressure measurement device.

6.5. An underwater instrumentation canister is a cylinder with ends which are circular plates that can be considered clamped at the outer edges, $r = a$ or $r = 1$. In order to design the ends, i.e., choose the thickness, h, for a given material system, one must determine the location and magnitude of the maximum stress when this canister is underwater with ambient pressure p_0.

Assume that the cylindrical portion introduces no in-plane loads to the ends. Find the maximum radial stress. Also what is the maximum circumferential stress? What elastic properties of the material are involved in finding the maximum radial stress? The maximum circumferential stress?

Figure 6.12. Underwater canister.

6.6. A flat circular plate roof is being designed to fit over an unused cave entrance. The outer radius is a, the thickness h, and the outer edge is considered clamped. If the weight density of the material used is ρ (lbs/in.^{3}), what is the maximum deflection, and the maximum stress in the plate due to gravity alone?

6.7. An air pump is constructed of a shaft, to which is clamped a disk of uniform thickness, at whose outer edge a soft gasket prevents air passage between the disk and the surrounding cylinder as shown on Figure 6.13.

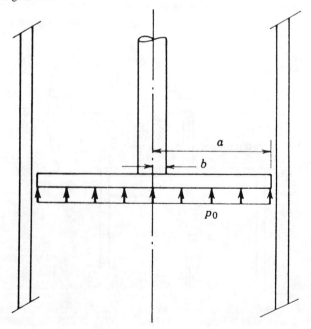

Figure 6.13. Air pump.

Assuming the disk is clamped to the shaft of radius b, and free at its outer edge of radius a, and a maximum differential pressure of p_0 (psi) can be built up, what is the expression for the maximum radial stress? What is the value of the maximum shear resultant, Q_r (lbs/in.)? Could Q_r have been determined in another way?

6.8. In a chemical plant, a certain process involving high gaseous, pressures requires a blow-out diaphragm which will blow at 100 psi pressure in order that expensive equipment will not be damaged. The flat diaphragm is 10" in radius, simply supported on its outer edge. Constructed of a brittle material with $v = 1/3$, and the ultimate tensile strength is 50000 psi. What thickness is required to have the plate fracture when the lateral pressure reaches 100 psi?

6.9. A circular plate of 14" diameter is used in a pressure system, and is composed of the steel of Problem 2. If the uniform lateral pressure is 50 psi, and the plate is simply supported at its outer edge, what is
   a. the thickness required?
   b. the maximum deflection?

6.10. A circular plate is used as a component in a pressure vessel. It has 14" diameter, is simply supported at the outer edge, and is composed of steel: $E = 30 \times 10^6$ psi, $\sigma_{all} = 40000$ psi, $v = 0.3$. Using maximum stress theory, what thickness, h, is required for a pressure differential of 50 psi?

6.11. What is the governing differential equation for the lateral deflection, w, for a circular plate subjected to a lateral pressure, $p(r)$, and a temperature distribution, $T(r, z)$?

6.12. A circular steel plate, used as a footing, rests on the ground and is subjected to a uniform lateral pressure, $p(r) = -p_0$ (psi). The ground deflects linearly with a spring constant of k (lbs/in./in.^{2}).
   a. What is the governing differential equation for this problem?
   b. What are the boundary conditions at the outer edge, $r = a$?

6.13. In Problem 6.7, write explicitly the four boundary condition equations involving the constants A, B, C, and E, etc. Do not bother to solve for the constants, however.

6.14. In Problem 6.8, the maximum radial and circumferential stresses are at the plate center and given by

$$\sigma_{r_{max}} = \sigma_{\theta_{max}} = \pm \tfrac{3}{8}(3 + v)\, \frac{p_0 a^2}{h^2} \quad \text{at} \quad z = \pm h/2 .$$

If the plate is 10" in radius, $v = 0.3$, $p_0 = 10$ psi, and $F_{ty} = 50000$ psi, what thickness is required if
   a. the maximum stress theory is used?
   b. the maximum distortion energy theory is used?
In each case assume the stress field is two dimensional, i.e., ignore σ_z.

Buckling of Columns and Plates

7.1. Derivation of the Plate Governing Equations for Buckling

The governing equations for a thin plate subjected to both in-plane and lateral loads have been derived previously. In those equations, there was one governing equation describing the relationship between the lateral deflection and the laterally distributed loading,

$$D\nabla^4 w = p(x, y)$$

and other equations dealing with in-plane displacements, related to in-plane loads

$$\nabla^4 u_0 = \nabla^4 v_0 = 0 .$$

As discussed previously, the equation involving lateral displacements and lateral loads is completely independent (uncoupled) from those involving the in-plane loadings and in-plane displacements.

However, it is true that when in-plane loads are compressive, upon attaining certain discrete values, these compressive loads do result in producing lateral displacements. Thus, there does occur a coupling between in-plane loads and lateral displacements, w. As a result, a more inclusive theory must be developed to account for this phenomenon, which is called *buckling* or *elastic instability*.

Unlike in developing the governing plate equations in Chapter 1, wherein the development began with the three dimensional equations of elasticity, the following shall begin with looking at the in-plane forces acting on a plate element, in which the forces are assumed to be functions of the midsurface coordinates x and y, as shown in Figure 7.1.

Looking now at the plate element of Figure 7.2 above, viewed from the midsurface in the positive y direction, the relationship between forces and displacements is seen, when the plate is subjected to both lateral and in-plane forces, i.e., when there is a lateral deflection, w (note obviously that in the figure the deflection is exaggerated).

Hence, the z component of the N_x loading per unit area is, for small slopes (i.e., the sine of the angle equals the angle itself in radians):

$$\frac{1}{dx\,dy}\left[\left(N_x + \frac{\partial N_x}{\partial x}\,dx\right)dy\left(\frac{\partial w}{\partial x} + \frac{\partial^2 w}{\partial x^2}\,dx\right) - N_x\,dy\,\frac{\partial w}{\partial x}\right]$$

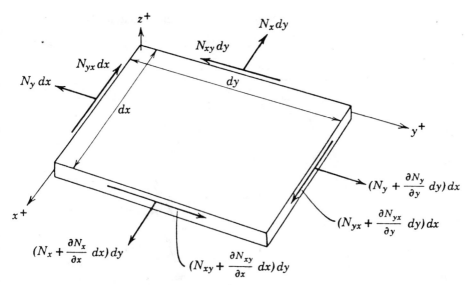

Figure 7.1. In-plane forces on a plate element.

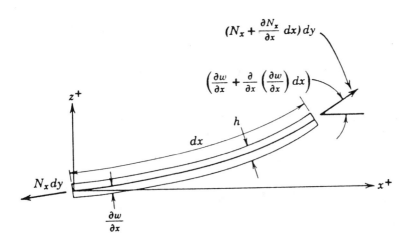

Figure 7.2. In-plane forces acting on a deflected plate element.

Neglecting terms of higher order, the force per unit planform area in the z direction is seen to be

$$N_x \frac{\partial^2 w}{\partial x^2} + \frac{\partial N_x}{\partial x} \frac{\partial w}{\partial x} \ . \tag{7.1}$$

Similarly, the z component of the N_y force per unit planform area is seen to be

$$N_y \frac{\partial^2 w}{\partial y^2} + \frac{\partial N_y}{\partial y} \frac{\partial w}{\partial y} \ . \tag{7.2}$$

Finally to investigate the z component of the in-plane resultants N_{xy} and N_{yx},

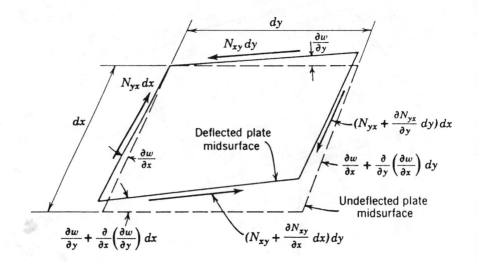

Figure 7.3. In-plane shear forces acting on a deflected plate element.

Hence, the z component per unit area of the in-plane shear resultants is:

$$\frac{1}{dx\,dy}\left\{\left(N_{xy} + \frac{\partial N_{xy}}{\partial x}\,dx\right)\left(\frac{\partial w}{\partial y} + \frac{\partial^2 w}{\partial x\,\partial y}\,dx\right)dy\right.$$

$$+ \left(N_{yx} + \frac{\partial N_{yx}}{\partial y}\,dy\right)\left(\frac{\partial w}{\partial x} + \frac{\partial^2}{\partial x\,\partial y}\,dy\right)dx$$

$$\left. - N_{xy}\,\frac{\partial w}{\partial y}\,dy - N_{yx}\,\frac{\partial w}{\partial x}\,dx\right\}.$$

Neglecting higher order terms, this results in

$$N_{xy}\,\frac{\partial^2 w}{\partial x\,\partial y} + \frac{\partial N_{xy}}{\partial x}\,\frac{\partial w}{\partial x} + N_{yx}\,\frac{\partial^2 w}{\partial x\,\partial y} + \frac{\partial N_{yx}}{\partial y}\,\frac{\partial w}{\partial x} \;. \tag{7.3}$$

With all the above z components of forces per unit area, the governing plate equation can be modified to include the effect of these in-plane forces on the governing plate equation.

$$D\nabla^4 w = p(x, y) + N_x \frac{\partial^2 w}{\partial x^2} + N_y \frac{\partial^2 w}{\partial y^2} + 2N_{xy} \frac{\partial^2 w}{\partial x \, \partial y}$$

$$+ \frac{\partial N_x}{\partial x} \frac{\partial w}{\partial x} + \frac{\partial N_y}{\partial y} \frac{\partial w}{\partial y}$$

$$+ \frac{\partial N_{xy}}{\partial x} \frac{\partial w}{\partial y} + \frac{\partial N_{yx}}{\partial y} \frac{\partial w}{\partial x} \ . \tag{7.4}$$

However, from in-plane force equilibrium, it is remembered from Equations (2.17) and (2.18), assuming no applied surface shear stresses, that

$$\frac{\partial N_x}{\partial x} + \frac{\partial N_{yx}}{\partial y} = 0 \tag{7.5}$$

$$\frac{\partial N_{xy}}{\partial x} + \frac{\partial N_y}{\partial y} = 0 \tag{7.6}$$

Substituting these into the expression above, the final form of the equation is found to be:

$$D\nabla^4 w = p(x, y) + N_x \frac{\partial^2 w}{\partial x^2} + N_y \frac{\partial^2 w}{\partial y^2} + 2N_{xy} \frac{\partial^2 w}{\partial x \, \partial y} \ . \tag{7.7}$$

Likewise, this governing plate equation can be reduced to the governing equation for a beam column by multiplying (7.7) by b (the width of the beam) and letting $\partial(\ )/\partial y = 0$, $v = 0$, $\overline{P} = -bN_x$ and $q(x) = bp(x)$, to provide

$$\frac{d^4 w}{dx^4} + k^2 \frac{d^2 w}{dx^2} = \frac{q(x)}{EI} \quad \text{where} \quad k^2 = \overline{P}/EI \ . \tag{7.8}$$

It should be noted that the load $\overline{P}$ defined above is an in-plane load which when positive produces compressive stresses, which differs from the convention used elsewhere throughout this text. However, it is commonly used in the literature on buckling, is convenient, so herein is denoted as a barred quantity.

7.2. Buckling of Columns Simply Supported at Each End

Solving Equation (7.8) by methods described previously, the solution can be written as:

$$w(x) = A \cos kx + B \sin kx + C + Ex + w_p(x) \tag{7.9}$$

where $w_p(x)$ is the particular solution for the loading $q(x)$. Consider, for example, the case wherein $q(x) = 0$, and the column is simply supported at each end. The boundary conditions, at $x = 0, L$, are then

$$w(0) = w(L) = 0 \tag{7.10}$$

$$M_x\left(\frac{L}{0}\right) = -EI\,\frac{\mathrm{d}^2 w}{\mathrm{d}x^2} = 0 \quad \text{or} \quad \frac{\mathrm{d}^2 w(0)}{\mathrm{d}x^2} = \frac{\mathrm{d}^2 w(L)}{\mathrm{d}x^2} = 0\,.$$

From the first boundary condition $A + C = 0$, from the third $A = 0$; hence, $C = 0$ also. From the second boundary condition $B \sin kL + EL = 0$, and from the fourth boundary condition

$$Bk^2 \sin kL = 0 = \frac{B\overline{P}}{EI}\,\sin kL = 0\,. \tag{7.11}$$

Note that in Equation (7.11) when $kL \neq n\pi$, then $B = E = 0$; when $kL = n\pi$, then $E = 0$, $B \neq 0$ but is indeterminate and

$$\overline{P} = n^2\pi^2\,\frac{EI}{L^2}\,. \tag{7.12}$$

It is thus seen that for most values of $\overline{P}$, the axial compressive loading, the lateral deflection w is zero ($A = B = C = E = 0$), and the in-plane and lateral forces and responses are uncoupled. However, for a countable infinity of discrete values of P, there is a lateral deflection, but it is of an indeterminate magnitude. Mathematically, this is referred to as an *eigenvalue problem* and the discrete values given in (7.12) are called *eigenvalues*. The resulting deflections, in this case

$$w(x) = B \sin kx$$

are called *eigenfunctions*.

The natural vibration of elastic bodies are also eigenvalue problems, where in that case the natural frequencies are the eigenvalues and the vibration modes are the eigenfunctions. This is treated in a subsequent chapter.

As to buckling, looking at Equation (7.12), as $\overline{P}$ increases, it is clear that the lowest buckling load occurs when $n = 1$, and at that particular load, the column will either inelastically deform and strain harden, or the column will fracture. Hence, $n < 1$ has no physical significance. The load

$$\overline{P} = \pi^2\,\frac{EI}{L^2} \tag{7.13}$$

is therefore the critical buckling load for this column for these boundary conditions. In this particular case the buckling load is called the Euler buckling load, since the Swiss mathematician was the first to solve the problem successfully.

Another way to phrase the buckling problem is exemplified by solving Equation 7.8, letting $q(x) = q_0 = $ constant. The resulting particular solution, in this case, is $q_0 = x^2/2P$. If the column is simply supported, solving the boundary value problem for the lateral deflection, results in

$$w(x) = \frac{q_0}{Pk^2 \sin kL}\left[\begin{array}{c} \cos kx\,\sin kL + \sin kL - \cos kL - \sin kL \\ -Lx\,\sin kL + k^2 x^2\,\sin kL \end{array}\right]\,. \tag{7.14}$$

In Equation (7.14), the solution of a boundary value problem, when the axial load $\overline{P}$ has values given in (7.12), $\sin kL = 0$ and $w(x)$ goes to infinity, or, more properly; since we have a small deflection linear mathematical model, $w(x)$ becomes indefinitely large.

Hence, whether we solve for the homogeneous solution of Equation (7.8), resulting in an eigenvalue problem, or we solve the nonhomogeneous Equation (7.8), resulting in a boundary value problem, the results are identical, when $\overline{P}$ has values given by (7.12), or physically where $\overline{P}$ attains the value given by (7.13), the column 'buckles'.

Note also that the buckling load, Equation (7.13), is not affected by any lateral load $q(x)$. The physical significance of a lateral load $q(x)$, however, is that the beam-column may deflect sufficiently, due to both the lateral and in-plane compressive loads, that the resulting curvature would cause bending stress which in addition to the compressive stresses may fracture or yield the column at a load less than or prior to attaining the buckling load.

These elastic stability considerations are very important in analyzing or designing any structure in which compressive stresses result from the loading, because in addition to insuring that the structure is not merely overstressed or overdeflected, in this case a new failure mode has been added, i.e., buckling.

7.3. Column Buckling with Other Boundary Conditions

From the previous section, the critical compressive buckling load $\overline{P}_{cr}$ is given as

$$\overline{P}_{cr} = \pi^2 \frac{EI}{L^2} \tag{7.15}$$

Numerous other texts derive critical buckling loads for columns with other boundary conditions, References 7.1 through 7.3.

For ease of use in analysis and design, but without derivations, the following column buckling equations are listed for the other classical boundary conditions.

Column with both ends clamped

$$\overline{P}_{cr} = 4\pi^2 \frac{EI}{L^2} \ . \tag{7.16}$$

Column with one end clamped and the other simply supported

$$\overline{P}_{cr} = \frac{\pi^2 EI}{(0.669L)^2} \ . \tag{7.17}$$

Column with one end clamped and the other end free

$$\overline{P}_{cr} = \frac{\pi^2 EI}{4L^2} \ . \tag{7.18}$$

7.4. Buckling of Plates Simply Supported on All Four Edges

Plate buckling qualitatively is analogous to column buckling, except that the mathematics is more complicated, and the conditions that result in the lowest eigenvalue (the actual buckling load) are not so lucid in many cases.

Whenever the in-plane forces are compressive, and are more than a few percent of the plate buckling loads (to be defined later), Equation (7.7) must be used rather than Equation (4.1) in the analysis of plates.

For the plate, just as the case of the beam-column, since the in-plane load that causes an elastic stability is not dependent upon a lateral load, to investigate the elastic stability we shall assume $p(x, y) = 0$ in Equation (7.7).

Consider, as an example, a simply supported plate subjected to constant in-plane loads N_x and N_y (let $N_{xy} = 0$), as shown in Figure 7.4.

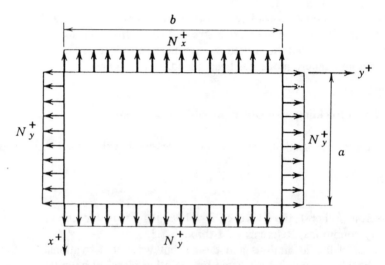

Figure 7.4. Rectangular plate subjected to in-plane loads.

Assume the solution of Equation (7.7) to be of the Navier form

$$w(x, y) = \sum_{m=1}^{\infty} \sum_{n=1}^{\infty} A_{mn} \sin \frac{m \pi x}{a} \sin \frac{n \pi y}{b} . \tag{7.19}$$

Substituting (7.19) into (7.7), it is convenient to define α to be

$$\alpha = N_y/N_x . \tag{7.20}$$

The solution to this eigenvalue problem is found to be

$$N_{x_{cr}} = -D\pi^2 \frac{\left[\left(\frac{m}{a}\right)^2 + \left(\frac{n}{b}\right)^2\right]^2}{\left[\left(\frac{m}{a}\right)^2 + \left(\frac{n}{b}\right)^2 \alpha\right]} . \tag{7.21}$$

Here the subscript cr denotes that this is a critical load situation – the plate buckles. Also note that N_x is a negative quantity, i.e., a load that causes compressive stresses.

Equation (7.21) is the complete set of eigenvalues for the simply supported plate, analogous to Equation (7.12) for the column. In other words for these discrete values of N_x and N_y, Equation (7.7) has nontrivial solutions wherein the lateral deflection is given by (7.15); for other values $w(x, y) = 0$.

Since we know that as the load increases, the plate will buckle at the lowest buckling load (or eigenvalue) and all the rest of the eigenvalues have no physical meaning, what values of m and n (the number of half sine waves) make N_x a minimum?

Defining the length to width ratio of the plate to be $r = a/b$ Equation (7.21) can be rewritten as

$$N_x = -\frac{D\pi^2}{a^2} \frac{[m^2 + n^2 r^2]^2}{[m^2 + n^2 r^2 \alpha]} . \tag{7.22}$$

Note if in Equation (7.22) $\alpha = 0$, $r = 1$, and $m = n = 1$, then

$$N_x = -\frac{4\pi^2 D}{a^2} . \tag{7.23}$$

Note the similarities between Equations (7.19) and (7.13).

The question remains; given a combination of N_x and N_y loadings, and a given

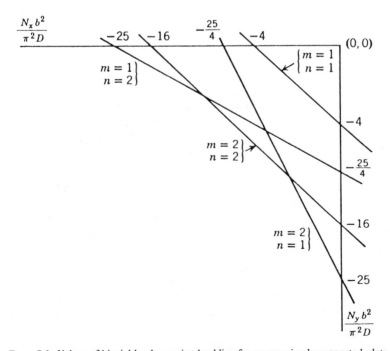

Figure 7.5. Values of biaxial loads causing buckling for square simply supported plate.

geometry r, what values of m and n provide the lowest buckling loads. One can make a plot such as Figure 7.5 below from manipulating Equation (7.22) (which is not shown to scale) for a square plate ($a = b$, $r = 1$).

It is seen from Figure 7.5 that for such a square plate, simply supported on all four edges, the plate will always buckle into a half sine wave ($m = n = 1$) under any combination of N_x and/or N_y, since that line is always closest to the origin, hence, the lowest buckling load situation.

Next consider a plate under an in-plane load in the x direction only, so $N_y = 0$, and $\alpha = 0$. In this case, Equation 7.21) can be written as

$$N_{x_{cr}} = -\frac{D\pi^2 a^2}{m^2}\left[\frac{m^2}{a^2} + \frac{n^2}{b^2}\right]^2 .$$ (7.24)

The loaded plate is shown in Figure 7.6.

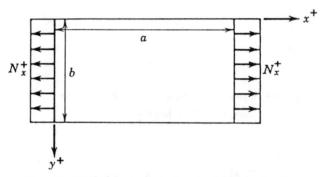

Figure 7.6. Plate subjected to in-plane load in the x-direction.

Examination of Equation (7.24) shows that the first term is merely the Euler column load (7.13) for a column of unit width, including Poisson ratio effects. The second term clearly shows the buckle resisting effect provided by the simply supported side edges, and this effect diminishes as the plate gets wider, i.e., as b increases. In fact as $b \to \infty$, (7.24) shows that the plate acts merely as an infinity of unit width beams, simply supported at the ends, and because they are 'joined together', the Poisson ratio effect occurs, i.e., D instead of EI appears.

It is obvious from Equation (7.24) that the minimum value of N_x occurs when $n = 1$, since n appears only in the numerator. Thus for an isotropic plate, simply supported on all four edges, subjected only to an uniaxial in-plane load the buckling mode given by (7.19) will always be one half sine wave [$\sin(y/b)$] across the span, regardless of the length or width of the plate.

Thus, since $n = 1$, Equation (7.24) can be written as

$$N_{x_{cr}} = -\frac{D\pi^2}{b^2}\left(\frac{m}{r} + \frac{r}{m}\right)^2$$ (7.25)

where it is remembered that $r = a/b$.

Note if $a < b$ (the plate wider than it is long), the second term is always less than the first, hence, the minimum value of N_x is always obtained by letting $m = 1$. Hence for $a \leqslant b$, the buckling mode for the simply supported plate is always

$$w(x, y) = A_{11} \sin\left(\frac{\pi x}{a}\right) \sin\left(\frac{\pi y}{b}\right). \tag{7.26}$$

In that case,

$$N_{x_{cr}} = -\frac{D\pi^2}{b^2}\left(\frac{1}{r} + r\right)^2. \tag{7.27}$$

To find out at what aspect ratio r, that N_x is truly a minimum, let

$$\frac{dN_{x_{cr}}}{dr} = 0 = -\frac{2D\pi^2}{b^2}\left(\frac{1}{r} + r\right)\left(-\frac{1}{r^2} + 1\right).$$

Therefore $r = 1$ provides that minimum value. Hence for $m = 1$, N_x is a minimum when $a = b$. Under that condition, from (7.27)

$$N_{x_{cr\,a=b}} = -\frac{4D\pi^2}{b^2} = -\frac{4D\pi^2}{a^2}. \tag{7.28}$$

Comparing this with the Euler buckling load of (7.13) for a simply supported column, it is seen that the continuity of a plate and the support along the sides of the plate provide a factor af at least 4 over the buckling of a series of strips (columns) that are neither continuous nor supported along the unloaded edges.

Now as the length to width ratio increases, as a/b increases, the buckling load (7.27) will increase, and one can ask, will $m = 1$ always result in a minimum buckling load, or is there another value of m which will provide a lower buckling load as r increases (i.e., $N_{x_{cr}}(m = 2) \leqslant N_{x_{cr}}(m = 1)$ for some value of r?)

Mathematically, this can be phrased as the following, using (7.25):

$$\left(\frac{m}{r} + \frac{r}{m}\right)^2 \overset{?}{\leqslant} \left(\frac{m-1}{r} + \frac{r}{m-1}\right)^2.$$

This states the condition under which the plate of aspect ratio r will buckle in m half sine waves in the loaded direction rather than $m - 1$ sine waves. Manipulating this inequality results in

$$m(m - 1) \leqslant r^2. \tag{7.29}$$

Equation (7.29) states that the plate will buckle in two half sine waves in the axial direction rather than one when $r \geqslant \sqrt{2}$. The plate will buckle in three half sine waves in the axial direction rather than 2, when $r \geqslant \sqrt{6}$, etc.

Again one can ask that when the plate buckles into $m = 2$ configuration, does a minimum buckling load occur, if so at what r and what is $N_{x_{cr\,(min)}}$?

From Equation (7.25)

$$\frac{\mathrm{d}N_{x_{cr}}}{\mathrm{d}r}(m=2)=0=-\frac{D\pi^2}{b^2}\,2\left(\frac{2}{r}+\frac{r}{2}\right)\left(-\frac{2}{r^2}+\frac{1}{2}\right)=0$$

or $r^2 = 4$, $r = 2$.

$$N_{x_{cr\,min}} = -\frac{4\pi^2 D}{b^2}\quad\text{for}\quad m=2\,.\tag{7.30}$$

This is the same value as is given in Equation (7.28) for $m = 1$. Proceeding with all values of r and m, the following graph can be drawn, which clearly shows the results (Figure 7.7).

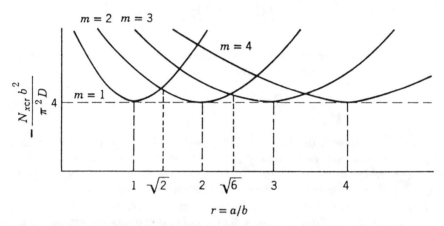

Figure 7.7. Buckling load as a function of aspect ratio for a simply supported plate.

Hence knowing the value of r, the figure provides the actual value of N_x and the corresponding value of the wave number m in the load direction. However, in practice for $r > 1$, universally one simply uses Equation (7.28) or (7.30) for the buckling load.

However, looking more closely at Equation (7.29), as m increases we see

$$m(m-1)\to m^2 = r^2\quad\text{or}\quad m = r = a/b\,.$$

This means that for long plates, the number of half sine waves of the buckles have lengths approximately equal to the plate width. Another way of stating it is that a long plate simply supported on all four edges and subjected to a uniaxial compressive load attempts to buckle into a number of square cells.

Remembering that $\sigma_x = N_x/h$, Equation (7.28) or (7.30) can be written as the following for $a/b \geqslant 1$,

$$\sigma_{cr} = -\frac{\pi^2 E}{3(1-v^2)}\left(\frac{h}{b}\right)^2\,.\tag{7.31}$$

7.5. Buckling of Plates with Other Loads and Boundary Conditions

The solution to the buckling of flat isotropic plates simply supported on all four sides subjected to uniaxial uniform compressive in-plate loads has been treated in detail. However, for many other boundary conditions, simple displacement functions like Equation (7.19) do not exist, and in some cases analytical, exact solutions analogous to Equations (7.21) and (7.31) have not been found. In those cases approximate solutions have been found using energy methods, which will be discussed in Chapter 9. These have been catalogued by Gerard and Becker (Reference 7.3) among

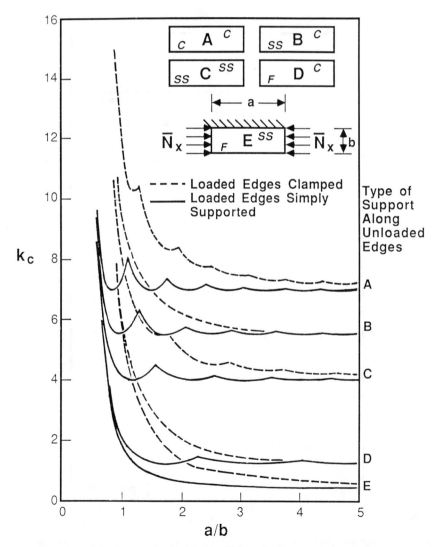

Figure 7.8. Compressive-buckling coefficients for flat rectangular plates.

others, and are presented in Figure 7.8 and k_c, given in the following equations:

$$\sigma_{x_{cr}} = -\frac{k_c \pi^2 E}{12(1 - v^2)}\left(\frac{h}{b}\right)^2 ; \quad N_{x_{cr}} = -\frac{k_c \pi^2 D}{b^2} \tag{7.32}$$

In many practical applications, the edge rotational restraints lie somewhere between fully clamped and simply supported along the unloaded edges. For the case of the loaded edges simply supported, the buckling coefficient, k_c, of Equation (7.32) are given by Gerard and Becker (Reference 7.3) as shown in Figure 7.9. The unloaded edge restraint, ε, is zero for simply supported edges and infinity for full clamping. Values in between these extremes require engineering judgment.

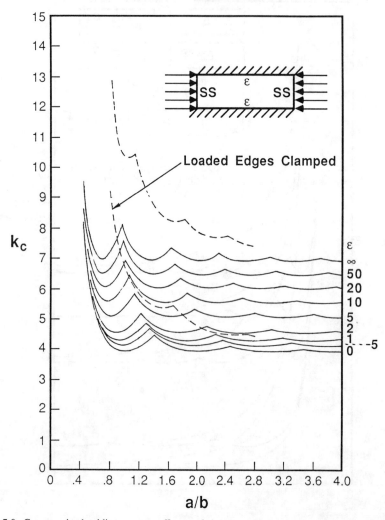

Figure 7.9. Compressive-buckling-stress coefficient of plates as a function of a/b for various amounts of edge rotational restraint.

For in-plane shear loading, the critical shear stress is given by the following equations:

$$\tau_{cr} = \frac{K_s \pi^2 E}{12(1 - v^2)} \left(\frac{h}{b}\right)^2 \ ; \quad N_{xy_{cr}} = \frac{K_s \pi^2 D}{b^2}$$

(7.33)

where K_s is given in Figure 7.10 for various boundary conditions (Reference 7.3).

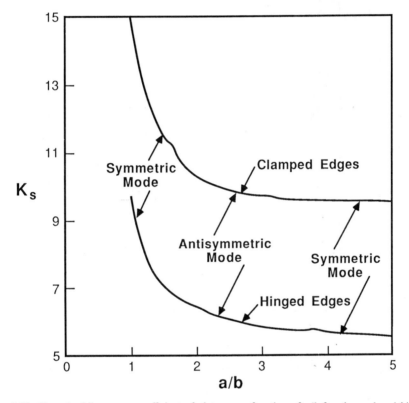

Figure 7.10. Shear-buckling-stress coefficient of plates as a function of a/b for clamped and hinged edges.

For rectangular plates subjected to in-plane bending loads, the following equation is used to determine the stress value for the buckling of the plate shown in Figure 7.11.

$$\sigma_B = \frac{k_b \pi^2 E}{12(1 - v^2)} \left(\frac{h}{b}\right)^2$$

(7.34)

where again ε is the value of the edge constraint as discussed previously.

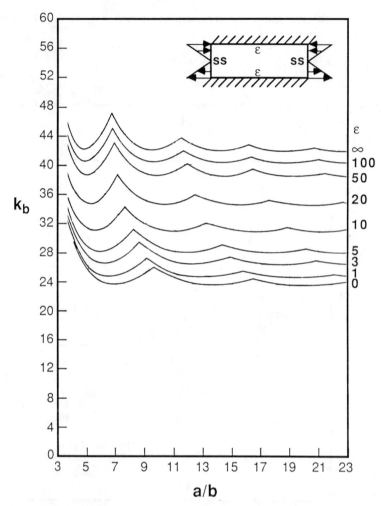

Figure 7.11. Bending-buckling coefficient of plates as a function of *a/b* for various amounts of edge rotational restraint.

7.6. References

7.1. Timoshenko, S. and J. Gere, *Elastic Stability*, McGraw-Hill Book Co., Inc., 2nd Edition, 1961.
7.2. Bleich, H. H., *Buckling of Metal Structures*, McGraw-Hill Book Co., Inc., 1952.
7.3. Gerard, G. and H. Becker, *Handbook of Structural Stability, Part 1 – Buckling of Flat Plates*, NACA TN 3781, 1957.

7.7. Problems

7.1. In a plate clamped on all four edges, $v = 0.25$ and loaded in the x direction the critical buckling stress is given by (from Reference 7.1)

$$\sigma_{cr} = -\frac{k\pi^2 D}{b^2 h} = -\frac{k\pi^2 E}{12(1-v^2)}\left(\frac{h}{b}\right)^2$$

where D is the flexural stiffness, b is the plate width, a is the plate length, and h is the plate thickness. k_c is given by

a/b	0.75	1.0	1.5	2.0	2.5	3.0
k_c	11.69	10.07	8.33	7.88	7.57	7.37

(a) Part of a support fixture for a missile launcher measures $45'' \times 15''$, and must support 145 000 lbs in axially compressive load. Its edges are all clamped. If the plate is composed of aluminum with $E = 10 \times 10^6$ psi, allowable 30 000 psi (both the tensile and compressive allowable stress is of magnitude 30 000 psi) and $v = 0.25$. What thickness is required to prevent buckling? What thickness is required to prevent overstressing?

(b) Suppose a steel plate of the same dimensions were used instead of the aluminum with the following properties: $E_{steel} = 30 \times 10^6$ psi, $v = 0.25$ and $\sigma_{allowable} = \pm 100\,000$ psi. What thickness is needed to prevent buckling? Will the steel plate be overstressed?

(c) The density of steel is 0.283 lbs/in³, the density of aluminum is 0.100 lbs/in³. Which plate will be lighter?

7.2. A structural component in the interior of an underwater structure consists of a square plate of dimension a, simply supported on all four sides. If the component is subjected to in-plane compressive loads in both the x and y directions of equal magnitude, find N_x.

7.3. An aluminum support structure consists of a rectangular plate simply supported on all four edges is subjected to an in-plane uniaxial compressive load. If the length of the plate in the load direction is 4 feet, the width 3 feet, determine the minimum plate thickness to insure that the plate would buckle in the elastic range, if the material properties are $E = 10 \times 10^6$ psi, $v = 0.3$ and the compressive yield stress, $\sigma_y = 30\,000$ psi.

7.4. A rectangular plate 4 feet × 2 feet is subjected to an inplane compressive load N_x in the longer direction as shown in Figure 7.6. How much weight of plate can be saved by using a plate clamped on all four edges rather than having the plate simply-supported on all four edges to resist the same compressive load $N_{x_{cr}}$? Express the answer as a percentage.

7.5. An aluminium plate measuring 6 feet × 3 feet, of thickness 0.1 inch is clamped on all four edges. Use the material properties in Problem 7.3 above.

(a) If it is subjected to a compressive in-plane load in the longer direction, what is the buckling stress?

(b) Now much higher is this buckling stress compared to the same plate simply supported on all four edges?

The Vibrations of Beams and Plates

8.1. Introduction

Through the previous chapter, the static behavior of beams, rods, columns, and plates has been treated to determine displacements and stresses. This is important because many structures are stiffness critical (maximum deflections are limited) or strength critical (maximum stresses are limited). In Chapter 7, the elastic stability of these structures was treated because that is a third way in which structures can be rendered useless. In most cases when a structure becomes elastically unstable, it cannot fulfill its structural purpose.

In this chapter, the vibration of beams and plates is studied in some detail. Many textbooks have been written dealing with this subject, but here, only an introduction is made to show how one approaches and deals with such problems.

In linear vibrations, both natural vibrations and forced vibrations are important. The former deals with natural characteristics of any elastic body, and these natural vibrations occur at discrete frequencies, depending on the geometry and material systems only. Such problems (like buckling) are *eigenvalue* problems, the natural frequencies are the *eigenvalues*, and the displacement field associated with each natural frequency are the *eigenfunctions*. One remembers that in a simple spring-mass system, there is one natural frequency and mode shape; in a system of two springs and two masses, there are two natural frequencies and two mode shapes. In a continuous elastic system, theoretically there are an infinite number of natural frequencies and a mode shape associated with each.

Forced vibrations occur when an elastic body is subjected to a time dependent force or forces. In that case the response to the forced virbations can be viewed as a linear superposition of all the eigenfunctions (vibration modes), each with an amplitude determined by the form of the forcing function. In forced vibrations, the forces can be cyclic (harmonic vibration) or non-cyclic, including shock loads (those which occur over very small times).

8.2. Natural Vibrations of Beams

Consider again the beam flexure equation discussed previously.

$$EI \frac{d^4w}{dx^4} = q(x). \tag{8.1}$$

It is seen that the forcing function $q(x)$ is written in terms of force per unit length. Using D'Alembert's Principle for vibration, an initial term can be written which is the mass times the acceleration per unit length. Also the forcing function can be a function of time, and of course the lateral deflections will be a function of both spatial and temporal coordinates. The result is that (8.1) becomes, for the flexural vibration,

$$EI \frac{\partial^4 w}{\partial x^4} = q(x, t) - \rho A \frac{\partial^2 w}{\partial t^2}. \tag{8.2}$$

In the above ρ is the *mass* density of the beam material, and A is the beam cross-sectional area, both of which are taken here as constants for simplicity.

As stated previously, natural vibrations are functions of the beam material properties and geometry only, and are inherent properties of the elastic body – independent of any load. Thus, for natural vibrations, $q(x, t)$ is set equal to zero, and (8.2) becomes

$$EI \frac{\partial^4 w}{\partial x^4} + \rho A \frac{\partial^2 w}{\partial t^2} = 0. \tag{8.3}$$

To solve this equation to obtain $w(x, t)$, in general, one can assume $w(x, t) = X(x)T(t)$, a separable solution, use separation of variables to obtain a spatial function $X(x)$ which satisfies all of the boundary conditions, and a harmonic function for $T(t)$, and thus arrive at a characteristic set of variables to satisfy (8.3) and its boundary conditions. In that process the natural frequencies and mode shapes are determined.

By way of a specific example, consider the beam to be simply supported at each end. Then the spatial function is a sine function such that

$$w(x, t) = \sum_{n=1}^{\infty} A_n \sin \frac{n\pi x}{L} \sin \omega_n t \tag{8.4}$$

where A_n is the amplitude, and ω_n is the natural circular frequency in radians per unit time for the nth vibrational mode.

Substituting (8.4) into (8.3) results in:

$$\sum_{n=1}^{\infty} A_n \left[\frac{n^4 \pi^4}{L^4} EI - \omega_n^2 \rho A \right] \sin \frac{n\pi L}{L} \sin \omega_n t = 0. \tag{8.5}$$

For this to be an equation, then for each value of n,

$$\omega_n = \frac{n^2 \pi^2}{L^2} \sqrt{\frac{EI}{\rho A}}. \tag{8.6}$$

It is seen that for each integer n, there is a different natural frequency, and from (8.4) a corresponding mode shape (i.e., $n = 1$, a one half sine wave; $n = 2$, two half sine waves, etc.).

Unlike in buckling where one looks for the lowest buckling load only, in vibrations each natural frequency is important, because if a beam were subjected to an oscillating load coinciding with any one natural frequency, little energy would be needed to cause the amplitude to grow until failure occurs.

The lowest natural frequency, $n = 1$ in this case is called the *fundamental* frequency. Theoretically n could increase to infinity. However, at some point the governing equation (8.3) does not apply, and thus the resulting frequencies given by (8.6) become meaningless. For a beam of an isotropic material the classical beam Equation (8.3) ceases to apply when the vibration half wave length approaches the beam depth, h, because then transverse shear deformation effects ($\varepsilon_{xz} \neq 0$) become important and (8.3) must be modified.

It is noted that for beams with boundary conditions other than simply supported at each end, the eigenfunctions (vibration modes) are not as simple as a sine wave. These are treated in detail in any of many fine texts on vibration. However, the natural frequencies, W_n, are catalogued below for use in analysis and design. In this case, Equation 8.6 is modified slightly for general use.

$$\omega_n = \alpha_n^2 \sqrt{\frac{EI}{\rho AL^4}} \tag{8.7}$$

where α_n^2 values are given by the following:

Cantilevered Beam: $\alpha_1^2 = 3.52,$ $\alpha_2^2 = 22.6,$ $\alpha_3^2 = 61.7$
Clamped-Clamped Beam: $\alpha_1^2 = 22.4,$ $\alpha_2^2 = 61.7,$ $\alpha_3^2 = 121.0$

8.3. Natural Vibrations of Plates

Consider again the equation for the bending of a rectangular plate subjected to a lateral load, $p(x, y)$, given by Equation (2.57).

$$D\nabla^4 w = p(x, y) . \tag{8.8}$$

If d'Alembert's Principle were used to accommodate the motion, one would add a term to the right hand side equal to the negative of the product of the mass per unit area and the acceleration in the z direction. In that case, the right hand side of (8.8) becomes:

$$p(x, y, t) - \rho h \frac{\partial^2 w}{\partial t^2} (x, y, t) \tag{8.9}$$

where both p and w are functions of time as well as space, ρ is the *mass* density of the material and h is the plate thickness. For forced vibration $p(x, y, t)$ causes the dynamic response, and can vary from a harmonic oscillation to an intense one time impact.

As discussed in the previous section, to study the natural vibrations $p(x, y, t)$ is set equal to zero, and the governing equation becomes the following homogeneous equation:

$$D\left[\frac{\partial^4 w}{\partial x^4} + 2\frac{\partial^4 w}{\partial x^2 \partial y^2} + \frac{\partial^4 w}{\partial y^4}\right] + \rho h \frac{\partial^2 w}{\partial t^2} = 0. \tag{8.10}$$

As done previously, one can assume a solution for the lateral deflection, which spatially satisfies the boundary condition, is harmonic in time, and satisfies (8.10) above. For the case of a plate simply-supported on all four edges, such a function is

$$w(x, y, t) = \sum_{m=1}^{\infty} \sum_{n=1}^{\infty} A_{mn} \sin\frac{m\pi x}{a} \sin\frac{n\pi y}{b} \sin\omega_n t \tag{8.11}$$

where a and b are the plate dimensions, A_{mn} is the vibration amplitude for each value of m and n, and ω_{mn} is the natural circular frequency in radians per unit time. Substitution of (8.11) into (8.10) results in

$$\omega_{mn} = \frac{\pi^2 D}{\sqrt{\rho h}}\left[\left(\frac{m^2}{a^2} + \frac{n^2}{b^2}\right)^2\right]^{1/2}. \tag{8.12}$$

In this case the fundamental natural frequency occurs for $m = n = 1$. Again, the amplitude A_{mn} cannot be determined from this linear eigenvalues problem, in which the eigenvalues are the natural frequencies of equation (8.12) and the corresponding eigenfunctions are the mode shapes, given in (8.11).

As in buckling, plates with other boundary conditions comprise more complicated problems, often most difficult to solve analytically. In many cases appropriate solutions are obtained using energy methods (Chapter 9).

8.4. Forced Vibrations of Beams and Plates

These will not be treated herein for two reasons. One is that it would extend the scope of this text past its intended purpose; and the other is because there are many fine texts at a basic level and voluminous literature dealing with the subject. Thomson (Reference 8.1) is such a text, and Leissa (Reference 8.2) provides solutions to numerous problems.

One fine paper by Dobyns (Reference 8.3) also given by Vinson and Sierakowski (Reference 8.4) provide solutions to the dynamic response of anisotropic plates subjected to a variety of impact loads of practical value. Those solutions are easily simplified to treat plates of isotropic materials.

8.5 References

8.1. Thomson, W. T., *Vibration Theory and Applications*, Englewood Cliffs: Prentice Hall Publishers, 1965.
8.2. Leissa, A. W., *Vibration of Plates*, National Aeronautics and Space Administration Special Publication, 1969.

8.3. Dobyns, A. L., The Analysis of Simply-Supported Orthotopic Plates Subjected to Static and Dynamic Loads, *AIAA Journal*, May 1981, pp. 642–650.
8.4. Vinson, J. R. and R. L. Sierakowski, *The Behavior of Structures Composed of Composite Materials*, Dordrecht: Martinus-Nijhoff Publishers, 1986.

8.6. Problems

8.1. A beam is 30 inches long, 1 inch wide and made of steel ($E = 30 \times 10^6$ psi, $v = 0.3$, weight density $\rho_w = 0,283$ lbs/in^3), simply supported at each end. What must the thickness h be to insure that the lowest natural frequency is not lower than 30 Hz?

8.2. For the beam of Problem 8.1 dimensions and material, what is the fundamental frequency if the beam is cantilevered?

8.3. For the beam of Problem 8.1 dimensions and material, what is the fundamental frequency if the beam is clamped at each end?

Energy Methods in Beams, Columns and Plates

9.1. Introduction

The use of energy principles provides an alternative to using the equilibrium equations, stress-strain relations and the strain-displacement equations for formulating the governing differential equations for elastic bodies such as beams, rods, columns, plates, and shells. However, the use of energy principles also provides the natural boundary conditions for the body – something not obtained from the previous approach. This advantage can be quite useful.

While the energy principles provide this alternative, they are of utmost use in obtaining *approximate solutions* to elasticity problems. Their utility is in finding easier, more rapid solutions, and obtaining solutions to problems which are very difficult because of their geometry, loading conditions, or their complexity through numerous components. Hence, when treating beam, plate, or shell problems of tapered or discontinuous thickness, numerous discontinuous loads, or a ring and stringer reinforced shell, energy principles provide great utility.

There are three energy principles used in structural mechanics: the Theorem of Minimum Potential Energy, the Theorem of Minimum Complementary Energy, and Reissner's Variational Theorem. However, Minimum Complementary Energy is less useful for the type of problems discussed in this text. Reissner's Variational Theorem is most useful in solving advanced problems which include transverse shear deformation. Hence, Minimum Potential Energy will be discussed herein.

9.2. Theorem of Minimum Potential Energy

For any generalized elastic body, the potential energy of that body can be written as follows:

$$V = \int_R W \, dR - \int_{S_T} T_i u_i \, dx - \int_R F_i u_i \, dR \qquad (9.1)$$

where: W = strain energy density function, defined in (9.4) below; R = volume of the elastic body; T_i = ith component of all surface tractions; u_i = ith component of all deformations; $F_i = i$th component of all body forces; and S_T = portion of the surfaces of the body over which surface tractions are prescribed.

In (9.1), the first term on the right hand side is the strain energy of the elastic body, usually expressed as U. The second and third terms are the work done by the surface tractions and the body forces, respectively. Body forces are those proportional to the mass of the body, and since homogeneous bodies are treated herein, are therefore proportioned to volume.

The Theorem of Minimum Potential Energy can be stated as: *Of all the displacements satisfying compatibility and the prescribed boundary conditions, those which satisfy equilibrium equations make the potential energy a minimum.* Thus in obtaining an approximate solution, compatibility is satisfied exactly and equilibrium is satisfied approximately.

To use the Theorem, the operation is simply that the variation of the potential energy is zero, i.e.,

$$\delta V = 0 . \tag{9.2}$$

Variation is analogous to partial differentiation herein. In what follows, only three operations are needed:

$$\frac{\mathrm{d}(\delta w)}{\mathrm{d}x} = \delta \left(\frac{\mathrm{d}w}{\mathrm{d}x} \right), \quad \delta(w^2) = 2w\delta w, \quad \int (\delta w)\, \mathrm{d}x = \delta \int w\, \mathrm{d}x . \tag{9.3}$$

In (9.1) the strain energy density function, W, is defined as follows in a Cartesian reference frame:

$$W = \tfrac{1}{2}\sigma_{ij}\varepsilon_{ij} = \tfrac{1}{2}\sigma_x\varepsilon_x + \tfrac{1}{2}\sigma_y\varepsilon_y + \tfrac{1}{2}\sigma_z\varepsilon_z$$
$$+ \sigma_{xy}\varepsilon_{xy} + \sigma_{xz}\varepsilon_{xz} + \sigma_{yz}\varepsilon_{yz} . \tag{9.4}$$

To utilize the Theorem of Minimum Potential Energy, it is necessary to define the stress-strain relations to replace the stresses in (9.4) by strains, and it is necessary to define strain-displacement relations to place the strains in terms of displacements.

9.3. Analysis of Beams Subjected to a Lateral Load

A simple first example of Minimum Potential Energy is a beam subjected to a distributed continuous lateral load, shown in Figure 9.1.

The beam is of length L, with b, and height h, subjected to a continuous lateral load $q(x)$ in force per unit length. The modulus of elasticity of the beam material is E, and the stress-strain relations and strain-displacement relations are, from previous descriptions:

$$\sigma_x = E\varepsilon_x, \quad \varepsilon_x = \frac{\mathrm{d}u}{\mathrm{d}x} = -z\,\frac{\mathrm{d}^2w}{\mathrm{d}x^2} . \tag{9.5}$$

From (9.4), (9.5), and remembering from the beam theory discussed earlier that

$$\sigma_y = \sigma_z = \sigma_{xy} = \varepsilon_{xz} = \varepsilon_{yz} = 0 , \tag{9.6}$$

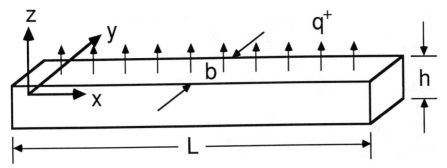

Figure 9.1. Beam subjected to a lateral load, *q(x)*.

the strain energy density function is seen to be:

$$W = \tfrac{1}{2}\sigma_x \varepsilon_x = \tfrac{1}{2}E\varepsilon_x^2 = \tfrac{1}{2}Ez^2\left(\frac{d^2w}{dx^2}\right)^2 . \tag{9.7}$$

Substituting (9.7) into (9.1), it is seen that

$$U = \int_0^L \int_{-h/2}^{+h/2} \int_{-h/2}^{+h/2} \frac{E}{2}\left(\frac{d^2w}{dx^2}\right)^2 z^2 \, dz \, dy \, dx = \frac{EI}{2} \int_0^L \left(\frac{d^2w}{dx^2}\right)^2 dx \tag{9.8}$$

where $I = bh^3/12$, for a rectangular cross-section.

Similarly, for the second term of (9.1), it is seen that

$$\int_{S_T} T_i u_i \, dS = \int_0^L q(x)w(x) \, dx . \tag{9.9}$$

Therefore, in the absence of body forces, (9.1) becomes

$$V = \frac{EI}{2} \int_0^L \left(\frac{d^2w}{dx^2}\right)^2 dx - \int_0^L q(x)w(x) \, dx . \tag{9.10}$$

Following (9.2), one obtains

$$\delta V = 0 = \frac{EI}{2} \int_0^L \delta\left(\frac{d^2w}{dx^2}\right)^2 dx - \int_0^L q(x)\delta w(x) \, dx . \tag{9.11}$$

In the above, it is seen that there is no variation in *EI* or *q(x)*, because they are specified. Integrating the first term by parts twice it is found that

$$\delta V = 0 = \left[EI \frac{d^2w}{dx^2} \delta\left(\frac{dw}{dx}\right) \right]_0^L - \left[EI \frac{d^3w}{dx^3} \delta w \right]_0^L$$

$$+ \int_0^L \left[EI \frac{d^4w}{dx^4} - q(x) \right] \delta w \, dx = 0 . \tag{9.12}$$

For this equation to be satisfied, the following must be true:

$$EI \frac{d^4w}{dx^4} = q(x) \tag{9.13}$$

and either

$$EI \frac{d^2w}{dx^2} = -M = 0 \quad \text{or} \quad \frac{dw}{dx} \quad \text{is specified,} \tag{9.14}$$

and either

$$EI_, \frac{d^3w}{dx^3} = -V = 0 \quad \text{or} \quad w \text{ is specified,} \tag{9.15}$$

at $x = 0$ and $x = L$, the ends of the beam.

Hence, using Minimum Potential Energy, the governing equation for the beam under transverse load is obtained, i.e., Equation 9.13. This product of the Variational Theorem is called generally the Euler–Lagrange equation.

In addition, Equations (9.14) and (9.15) are the natural boundary conditions for the beam. Notice that they include the classical boundary conditions of simple support, clamped and free edges. These are a by-product of employing the energy principle.

However, an equally useful way to employ the Theorem of Minimum Potential Energy is through assuming a function for the dependent variable, in this case $w(x)$, which is a reasonable function for that variable, i.e., one which is continuous, single valued and satisfies the boundary conditions.

As an example, consider a beam simply supported at each end subjected to a lateral load $q(x)$. In this case a reasonable function to assume is

$$w(x) = A \sin \frac{\pi x}{L} \tag{9.16}$$

Substituting (9.16) into (9.10) results in the following

$$V = \frac{EI}{2} \int_0^L A^2 \frac{\pi^4}{L^4} \sin^2 \frac{\pi x}{L} \, dx - \int_0^L q(x) A \sin \frac{\pi x}{L} \, dx =$$

$$= \frac{EI}{2} A^2 \frac{\pi^4}{L^4} \left(\frac{L}{2}\right) - A \int_0^L q(x) \sin \frac{\pi x}{L} \, dx. \tag{9.17}$$

In the variation process, here the variable is A, hence,

$$\delta V = 0 = 2A \, \delta A \, \pi^4 \frac{EI}{4L^3} - \delta A \int_0^L q(x) \sin \frac{\pi x}{L} \, dx \tag{9.18}$$

and the Euler–Lagrange Equation here is the following:

$$A = \frac{2L^3}{\pi^4 EI} \int_0^L q(x) \sin \frac{\pi x}{L} \, dx. \tag{9.19}$$

From (9.16) and (9.19), all displacements and stresses are calculated.

The above method can be generalized in that one could assume for the displacement (9.16), the following

$$w(x) = \sum_{n=1}^{\infty} A_n \sin \frac{n\pi x}{L} \tag{9.20}$$

each term of which satisfies the simple-support boundary conditions. In that case, for the Euler–Lagrange equation (9.19), one obtains an $N \times N$ matrix of algebraic equations to obtain the $A_i (i = 1, \ldots, N)$. Obviously if N equals infinity the exact solution is obtained, but very accurate approximate solutions are obtained for low values of N, but the number of terms required for accuracy depends on the load function, $q(x)$.

9.4. The Buckling of Columns

In this case the strain energy is again given by Equation 9.8, where neglecting body forces F_i, the work done by surface tractions is given as follows:

$$\int_{S_T} T_i u_i \, dS = -\int_0^L \left\{ P \left[\frac{du_0}{dx} + \frac{1}{2} \left(\frac{dw}{dx} \right)^2 \right] \right\} dx .$$

This equation incorporates the more comprehensive theory employed in Chapter 7 to include buckling, and as discussed previously, to calculate buckling loads, $u_0 = 0$, because at incipient buckling the arc length of the buckled column is equal to the original length. Also, in the above, P is a tensile load, considered constant to make the problem linear. Therefore,

$$V = \frac{EI}{2} \int_0^L \left(\frac{d^2 w}{dx^2} \right)^2 dx + \frac{P}{2} \int_0^L \left(\frac{dw}{dx} \right)^2 dx . \tag{9.21}$$

Taking the variation of the potential energy, one obtains the following Euler–Lagrange equation as in Equation (9.13)

$$EI \frac{d^4 w}{dx^4} - P \frac{d^2 w}{dx^2} = 0 \tag{9.22}$$

as well as the natural boundary conditions discussed previously. However, assuming a form of $w(x)$, which satisfies the boundary conditions for the column, which approximates the exact buckled shape will provide an approximation to the exact buckling load.

Consider a column simply supported at each end, if one uses (9.16) in (9.21) and takes variation of A, the result is:

$$P = -\pi^2 \frac{EI}{L^2} . \tag{9.23}$$

It is seen that this is the exact buckling load, because the exact buckling mode (9.16) was utilized. Some other approximate displacement functions satisfying the boundary conditions would give an approximate buckling load. It can be proven that such an approximate buckling load will always be greater than the exact buckling load. However, as long as the assumed displacement satisfies the boundary conditions, the error is never more than a very few percent of the exact value.

9.5. Vibration of Beams

The energy principle to utilize in dynamic analysis is Hamilton's Principle which employs the functional

$$I = \int_{t_1}^{t_2} (T - V)\, dt.$$ (9.24)

Hamilton's Principle states that in a conservative system

$$\delta I = 0.$$ (9.25)

In the above, the potential energy, V, is given by Equation (9.1), and T is the kinetic energy of the body. In a beam undergoing flexural vibration, the kinetic energy would be

$$T = \frac{1}{2} \int_0^L \rho A \left(\frac{\partial w}{\partial t}\right)^2 dx$$ (9.26)

where ρ is the *mass* density of the material, A is the beam cross-sectional area, and obviously $\partial w/\partial t$ is the velocity of the beam.

Using Hamilton's Principle in the same way that was done before for Minimum Potential Energy, the resulting Euler–Lagrange equation is

$$EI \frac{\partial^4 w}{\partial x^4} + \rho h \frac{\partial^2 w}{\partial t^2} = 0$$ (9.27)

which is identical to Equation 8.3. Also resulting are the natural boundary conditions, discussed previously.

Considering a beam simply supported at each end, if Equation (9.20) is modified to include a harmonic motion with time, as

$$w(x, t) = A \sin \frac{n\pi x}{L} \cos \omega_n t$$

the result is an Euler–Lagrange equation of

$$\omega_n = \frac{n^2 \pi^2}{L^2} \sqrt{\frac{EI}{\rho A}}$$ (9.28)

which is the exact solution (see Equation 8.6) because the exact mode shape was assumed. Again, if the assumed displacement function is approximate, then approximate natural frequencies will be obtained; are higher than the exact frequencies, but the error will be at most a few percent.

Note that in assuming mode shape functions in both buckling and vibration problems (eigenvalue problems), the closer the assumed function is to the exact mode shape, the lower the resulting eigenvalue will be, and of course it will be closer to the exact eigenvalue.

9.6. Minimum Potential Energy for Rectangular Plates

The strain energy density function, W, for a three dimensional solid in rectangular coordinates is given by Equation (9.4). The assumptions associated with the classical plate theory of Chapter 2 are employed to modify (9.4) for a rectangular plate. If transverse shear deformation is neglected, then $\varepsilon_{xz} = \varepsilon_{yz} = 0$. If there is no plate thickening, then $\varepsilon_z = 0$. From Equations (1.9), (1.10), and (1.12), stresses are written in terms of strains, such that for the classical plate,

$$\sigma_x = \frac{E}{(1 - v^2)} [\varepsilon_x + v\varepsilon_t] ; \quad \sigma_y = \frac{E}{(1 - v^2)} [\varepsilon_y + v\varepsilon_x] ; \quad \sigma_{xy} = \frac{E}{(1 + v)} \varepsilon_{xy} .$$

(9.29)

Therefore, (9.4) becomes

$$W = \frac{E\varepsilon_x}{2(1 - v^2)} (\varepsilon_x + v\varepsilon_y) + \frac{E\varepsilon_y}{2(1 - v^2)} (\varepsilon_y + v\varepsilon_x) + \frac{E}{(1 + v)} \varepsilon_{xy}^2 .$$

(9.30)

If the plate is subjected to bending and stretching, the deflection functions are given by Equations (2.24) through (2.28). Substituting these into (9.30) results in the following:

$$W = \frac{1}{2} \frac{E}{(1 - v^2)} \left\{ \left(\frac{\partial u_0}{\partial x}\right)^2 + \left(\frac{\partial v_0}{\partial y}\right)^2 + 2v \left(\frac{\partial u_0}{\partial x}\right)\left(\frac{\partial v_0}{\partial y}\right) + \frac{1 + v}{2} \left[\frac{\partial u_0}{\partial y} + \frac{\partial v_0}{\partial x}\right]^2 \right\}$$

$$+ \frac{Ez^2}{2(1 - v^2)} \left[\left(\frac{\partial^2 w}{\partial x^2}\right)^2 + \left(\frac{\partial^2 w}{\partial y^2}\right)^2 + 2v \left(\frac{\partial^2 w}{\partial x^2}\right)\left(\frac{\partial^2 w}{\partial y^2}\right) \right]$$

$$+ \frac{Ez^2}{(1 + v)} \left(\frac{\partial^2 w}{\partial x \, \partial y}\right)^2 .$$

(9.31)

From this the strain energy $U (= \int_R W \, dR)$ is found.

$$U = \frac{K}{2} \int_0^a \int_0^b \left\{ \left(\frac{\partial u_0}{\partial x} + \frac{\partial v_0}{\partial y}\right)^2 - 2(1 - v) \frac{\partial u_0}{\partial x} \frac{\partial v_0}{\partial y} + \frac{1 - v}{2} \left(\frac{\partial u_0}{\partial y} + \frac{\partial v_0}{\partial x}\right)^2 \right\} dx \, dy$$

$$+ \frac{D}{2} \int_0^a \int_0^b \left\{ \left(\frac{\partial^2 w}{\partial x^2} + \frac{\partial^2 w}{\partial y^2}\right)^2 - 2(1 - v) \left[\left(\frac{\partial^2 w}{\partial x^2}\right)\left(\frac{\partial^2 w}{\partial y^2}\right) - \left(\frac{\partial^2 w}{\partial x \, \partial y}\right)^2 \right] \right\} dx \, dy .$$

(9.32)

It is seen that the first term is the extensional or in-plane strain energy of the plate, and the second is the bending strain energy of the plate. In the latter, it is seen that the first term is proportional to the square of the average plate curvature, while the second term is known as the Gaussian curvature.

For the plate the total work term due to surface traction is seem to be

$$
\int_{S_T} T_i u_i \, dS = \int_0^1 \int_0^b p(x, y) \, dx \, dy - \int_0^a \int_0^b \left\{ N_x \left[\frac{\partial u_0}{\partial x} + \frac{1}{2} \left(\frac{\partial w}{\partial x} \right)^2 \right] \right.
$$
$$
\left. + N_y \left[\frac{\partial v_0}{\partial y} + \frac{1}{2} \left(\frac{\partial w}{\partial y} \right)^2 \right] + N_{xy} \left[\left(\frac{\partial u_0}{\partial y} + \frac{\partial v_0}{\partial x} \right) + \left(\frac{\partial w}{\partial x} \right) \left(\frac{\partial w}{\partial y} \right) \right] \right\} \, dx \, dy .
$$

$$(9.33)$$

Hence, in (9.32) and (9.33) if one considers a plate subjected only to a lateral load $p(x, y)$, one assumes $u_0 = v_0 = N_x = N_y = N_{xy} = 0$. If one is considering in-plane loads only (except for buckling) assume $w(x, y) = p(x, y) = 0$. If one is looking for buckling loads, assume $p(x, y) = u_0 = v_0 = 0$. The rationale for all of this has been discussed previously.

9.7. The Buckling of a Plate under Unaxial Load, Simply Supported on Three Sides, and Free on an Unloaded Edge

The most beneficial use of the Minimum Potential Energy Theorem occurs when one cannot formulate a suitable set of governing differential equations, when one cannot guess the deformation pattern, and/or when one cannot ascertain a consistent set of boundary conditions. In that case one can make a reasonable assumption of the displacements, and then solves for an approximate solution. This is illustrated in the following.

Consider the plate shown below.

The governing differential equation for this is obtained from Equation (7.7).

$$
\frac{\partial^4 w}{\partial x^4} + 2 \frac{\partial^4 w}{\partial x^2 \, \partial y^2} + \frac{\partial^4 w}{\partial y^4} = \frac{N_x}{D} \frac{\partial^2 w}{\partial x^2} .
$$

$$(9.34)$$

To solve for the buckling load directly, a Levy type solution may be assumed:

$$
w(x, y) = \sum_{m=1}^{\infty} \psi_m(y) \sin \frac{m \pi x}{a} .
$$

$$(9.35)$$

Substituting (9.35) into (9.34) results in the following ordinary differential equation to solve:

$$
\lambda_m^4 \psi - 2 \lambda_m^2 \psi'' + \psi^{iv} = - \frac{N_x}{D} \lambda_m^2 \psi
$$

$$(9.36)$$

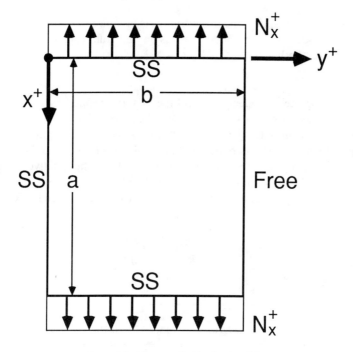

Figure 9.2. Plate studied in Section 9.7.

where

$$\lambda_m = \frac{m\pi}{a}, \quad ()'' = \frac{d^2()}{dy^2}, \quad \text{and} \quad ()^{iv} = \frac{d^4()}{dy^4}$$

Letting $\overline{N}_x = -N_x$, Equation (9.36) can be solved with the result that

$$\psi_m(y) = A\cosh\alpha y + B\sinh\alpha y + C\cos\beta y + E\sin\beta y \tag{9.37}$$

where

$$\alpha = \left[\lambda_m^2 + \lambda_m\sqrt{\frac{\overline{N}_{xm}}{D}}\right]^{1/2}$$

$$\beta = \left[-\lambda_m^2 + \lambda_m\sqrt{\frac{\overline{N}_{xm}}{D}}\right]^{1/2}.$$

The boundary conditions on the $y = 0$ and b edges are

$$\begin{aligned}
w(x, 0) &= 0 &\rightarrow& \quad \psi(0) = 0 \\
M_y(x, 0) &= 0 &\rightarrow& \quad \psi''(0) = 0 \\
M_y(x, b) &= 0 &\rightarrow& \quad \psi''(b) - v\lambda_m^2\psi(b) = 0 \\
V(x, b) &= 0 &\rightarrow& \quad \psi'''(b) - (2 - v)\lambda_m^3\psi'(b) = 0.
\end{aligned} \tag{9.38}$$

It is clear that the first two boundary conditions require that $A = C = 0$. Satisfying the other two boundary conditions results in the following relationship for the eigenvalues (i.e., the buckling load $\overline{N}_x = -N_x$).

$$-\beta \tanh \alpha b [\alpha^2 - v\lambda_m^2]^2 + \alpha \tan \beta b [\beta^2 + v\lambda_m^2]^2 = 0. \tag{9.39}$$

Thus, knowing the plate geometry and the material properties, one can solve for the buckling loads for each value of m. It can be shown that the minimum buckling load will occur for $m = 1$, thus a one-half sine wave in the longitudinal direction. However, note the complexity both in obtaining Equation (9.39), and then using that equation to obtain the buckling load, compared to the relative simplicity of Section 7.4 for solving the simpler problem of the plate completely simple supported on all four edges. The solutions of this problem have been catalogued in Reference 7.1 and are given below:

$$N_x = -\frac{k\pi^2 D}{b^2} \quad \text{and} \quad \sigma_{cr} = -\frac{k\pi^2 E}{12(1-v^2)} \left(\frac{h}{b}\right)^2.$$

For $v = 0.25$

a/b	0.50	1.0	2.0	3.0	4.0	5.0
k	4.40	1.44	0.698	0.564	0.516	0.506

Now to solve the same problem using Minimum Potential Energy. However, before doing so a brief discussion regarding boundary conditions is in order. They can be divided into two categories: geometric and stress. Geometric boundary conditions involve specifications on the displacement function and the first derivative, such as specifying the lateral displacement w or the slope at the boundary, $\partial w/\partial x$ or $\partial w/\partial y$, stress boundary conditions involve the specifications of the second and third derivative of the displacement function, such as the stress couples, M_x, M_y, M_{xy}, or the transverse shear resultants Q_x, Q_y, or the effective transverse shear resultants V_x, V_y discussed in Chapter 2.

In using the Minimum Potential Energy Theorem, one must choose a deflection function that *at least* satisfies the geometric boundary conditions specified on the boundaries. This suitable function will give a reasonable approximate solution. Better yet, by assuming a deflection function that satisfies all specified boundary conditions, one can achieve a very good approximate solution. If one could choose a deflection function that satisfies all boundary conditions and the governing differential equation for the problem also, this is the exact solution! Finally, if one chose a deflection function that did not satisfy even the geometric boundary conditions, the solution would be inaccurate because in effect the solution would not be for the problem to be solved, but for some other problem for which the assumed deflection does satisfy the geometric boundary conditions.

In this example, the following function is assumed for the lateral deflection:

$$w(x, y) = Ay \sin \frac{\pi x}{a} \tag{9.40}$$

This satisfies all boundary conditions on the $x = 0, a$ edges. It satisfies the geometric boundary condition that $w(x, 0) = 0$, but does not satisfy the stress boundary conditions that $M(x, 0) = M(x, b) = V(x, b) = 0$. Substituting Equation (9.40) and its derivatives into Equations (9.32) and (9.33), where of course $N_y = N_{xy} = p(x, y) = 0$ produces

$$V = \frac{D}{2} \int_0^a \int_0^b \left\{ \left[-Ay \frac{\pi^2}{a^2} \sin \frac{\pi x}{a} \right]^2 + 2(1 - v) \left[-A \frac{\pi}{a} \cos \frac{\pi x}{a} \right]^2 \right\} dx\, dy$$

$$+ \frac{N_x}{2} \int_0^a \int_0^b A^2 y^2 \frac{\pi^2}{a^2} \cos^2 \frac{\pi x}{a}\, dx\, dy. \tag{9.41}$$

Integration Equation (9.41) gives

$$V = A^2 D \left[\frac{\pi^4}{a^3} \frac{b^3}{3} + 2(1 - v) \frac{\pi^2 b}{a} \right] + N_x A^2 \frac{\pi^2 b^3}{3a}.$$

Setting $\delta V = 0$, where the only variable with which to take a variation is A, produces the requirement that

$$N_x = - \left[\frac{\pi^2 D}{a^2} + \frac{6(1 - v)D}{b^2} \right]. \tag{9.42}$$

To compare this approximate result with the exact solution shown previously, let $a/b = 1$, and $v = 0.25$. From Equation (9.42)

$$N_{x_{cr}} = - 1.456 \frac{\pi^2 D}{b^2}. \tag{9.43}$$

In the exact solution, the coefficient is 1.440. Hence, the difference between the approximate solution and the exact solution is approximately 1%, yet the three stress boundary conditions on the $y =$ constant edges were not satisfied.

9.8. Functions to Assume in the Use of Minimum Potential Energy for Solving Beam, Column, and Plate Problems

In the use of Minimum Potential Energy methods to solve beam, column, and plate problems, one usually needs to assume an expression for the lateral deflection $w(x)$ for the beam or column, and $w(x, y)$ for the plate. These must be single valued, continuous functions that satisfy all the boundary conditions, or at least the geometric ones. Below are a few functions useful in the solutions of beam and column problems. Simple-simple

$$w(x) = \sum_{n=1}^{\infty} A_n \sin \frac{n \pi x}{L} \tag{9.44}$$

Simple-free

$$w(x) = Ax \tag{9.45}$$

Clamped-clamped

$$w(x) = A\left[1 - \cos\frac{2m\pi x}{a}\right] \tag{9.46}$$

Clamped-free

$$w(x) = Ax^2 \tag{9.47}$$

Clamped-simple

$$w(x) = A[L^3 x - 3Lx^3 + 2x^4] \tag{9.48}$$

Free-free

$$w = A. \tag{9.49}$$

In the case of a plate with varied boundary conditions, let $w(x, y) = f(x)g(y)$ where for $f(x)$ and $g(y)$ use the appropriate beam functions above. For example, consider a plate clamped on edges $y = 0$ and $y = b$, and clamped at $x = 0$ and simply supported at $x = a$. Assume the function:

$$w(x, y) = A_m[L^3 x - 3Lx^3 + 2x^4]\left[1 - \cos\frac{2m\pi y}{b}\right]. \tag{9.50}$$

Keep in mind none of the above functions is unique, and thus the engineer may use his ingenuity to conceive functions best for the solution of that particular problem. For instance, suppose a plate had one edge simply supported at $y = 0$, $0 \leqslant x \leqslant a/2$, and clamped from $a/2 \leqslant x \leqslant a$. No analytical solution could be obtained but an approximate solution using energy methods is always attainable.

9.9. Problems

9.1. Consider a steel plate ($E = 30 \times 10^6$ psi, $v = 0.25$, $\sigma_y = 30\,000$ psi) used as a portion of a bulkhead on a ship. The bulkhead is 60″ long and 30″ wide subjected to an in-plane compressive load in the longer direction. What thickness must the plate be to have a buckling stress equal to the yield stress if:
(a) the plate is simply supported on all four edges?
(b) the plate is simply supported on three edges and free on one unloaded edge?

9.2. Given a column of width b, height h, and length L, simply supported at each end, use the principle of Minimum Potential Energy to determine the buckling load, if one assumes the deflection to be

(a) $w(x) = A\dfrac{x}{L}(L - x)$

(a) $w(x) = \dfrac{A}{L^3}[2Lx^3 - x^4 - L^3 x]$

where in each case A is an amplitude.

Do the deflections assumed above satisfy the geometric boundary conditions? Do they satisfy the stress boundary conditions?

9.3. Consider the plates below, each subjected to a uniform axial compressive load per inch of width, $\overline{N}_x = -N_x$ (lbs./in.) in the x direction. Determine a suitable deflection function $w(x, y)$ for each case for subsequent use in the Principle of Minimum Potential Energy to determine the critical load $\overline{N}_y$.

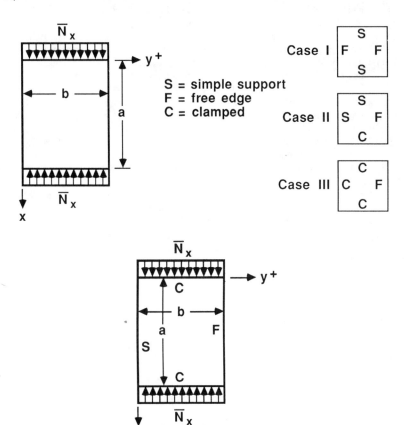

9.4. For an end plate in a support structure with the following boundary conditions, use the Principle of Minimum Potential Energy to determine the buckling load, if one assumes the deflection function to be $w = A_y[1 - \cos(2\pi x/a)]$.

9.5. Consider a rectangular plate of $0 \leqslant x \leqslant a$, $0 \leqslant y \leqslant b$, $-h/2 \leqslant z \leqslant h/2$. If the lateral deflection $w(x, y)$ is assumed to be in a separable form $w = f(x)g(y)$, and if $w = 0$ on all boundaries, determine the amount of strain energy due to the terms comprising the Gaussian curvature.

9.6. The base of a missile lauch platform consists in part of vertical rectangular plates of height a, and width b, where $a > b$. They are tied into the foundation below and the platform above such that those edges are considered clamped. However, on their vertical edges they are tied into I beams, such that those edges can only be considered simply supported. Using the Theory of Minimum Potential Energy, derive the equation for the buckling load per inch of edge distance, $N_{x_{cr}}$, for these plates, using a suitable deflection function, so that the plates can be designed to resist buckling.

9.7. An alternative to the design of Problem 9.6 would be to 'beef up' the vertical support beams such that the plate members can be considered to have their vertical edges clamped. Thus the plates have all four edges clamped. Employing a suitable deflection function, use the Theorem of Minimum Potential Energy to determine an expression for the critical buckling load per unit edge distance, $N_{x_{cr}}$, to use in designing the plates. Is the plate with all edges clamped thicker or thinner than the one with the sides simply supported by Problem 9.6, to have the same buckling load?

9.8.   The legs of a water tower consist of three columns of length a, constant flexural stiffness EI, simply supported at one end and clamped at the other end. Using the Theorem of Minimum Potential Energy, and a suitable function for the lateral deflection, calculate the buckling load P_{cr} for each leg, in order that they may be properly designed.

9.9.   Consider a beam of length L, and constant cross-section, i.e., EI is a constant. The beam is subjected to a load $q(x) = a + c(x/L)$, (lbs./in.) applied laterally where a and c are constants. The beam is simply supported on both ends. Using Minimum Potential Energy, and assuming $w(x) = B \sin(\pi x/L)$, determine the maximum deflection, w, and the maximum bending stress, σ_x. Consider the beam to be of unit width, i.e., $b = 1$.

9.10. A beam of length L, and constant cross-section (EI = constant) is subjected to a lateral load $q(x) = (q_0 x)/L$, where q_0 is a constant, and is simply supported at each end. Using Minimum Potential Energy, and assuming $w(x) = A \sin(\pi x/L)$, where A is a constant to be determined, determine the maximum deflection, w, and the maximum stress, σ_x, in the beam.

9.11. Consider the beam of Section 9.3 to be simply supported at each end and subjected to a uniform lateral load q_0 (lbs./in.). Assuming the deflection to be $W(x) = A \sin(\pi x/L)$, use the Principle of Minimum Potential Energy to determine A.

9.12. Consider a beam-column simply supported at one end and clamped at the other. Using the Theorem of Minimum Potential Energy, and assuming an admissible form for the lateral deflection, $w(x)$, calculate the in-plane load, P_{cr} (lbs.), to buckle the column.

9.13. Consider a beam of stiffness EI, length L, width b, height h, simply supported at each end, subjected to a uniform lateral load, q_0 (lb/in). Use Minimum Potential Energy, employing a deflection function

$$w(x) = \sum_{n=1}^{N} A_n \sin \frac{n \pi x}{L}$$

where $N = 3$, to determine the maximum deflection and maximum stress. Compare the answer with the exact solution.

9.14   Consider a column of length L, clamped at one end and simply supported at the other end. Using a buckling mode shape of

$$w(x) = A[L^3 x - 3Lx^3 + 2x^4]$$

where A is the buckle amplitude. Use Minimum Potential Energy to determine the axial critical buckling load, P_{cr}.

9.15. Consider a beam of constant flexural stiffness EI, of length L, clamped at each end. Using Hamilton's Principle, and an assumed deflection of

$$w(x, t) = A[1 - \cos(2\pi x/L)] \sin \omega_n t,$$

determine the fundamental natural frequency, and compare it with the exact solution.

Cylindrical Shells

There are entire texts dealing with shell theory, References 10.1 through 10.15. Most begin with a chapter on topology, because the natural coordinate systems for shells are curvilinear, and in general are fairly complicated to work with. Since this is an introductory text, such involved, lengthy considerations are not included. Only cylindrical shells will be treated because of their relative simplicity and because of their wide usage. In fact, the derivation of the governing differential equations is omitted, since to do it properly would require involving general topics of generalized curvilinear coordinate systems and topology.

10.1. Cylindrical Shells under General Loads

The simplest of all shell geometrics is that of the circular cylindrical shell shown below:

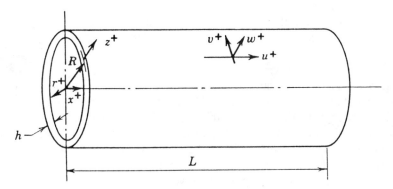

Figure 10.1. Circular cylindrical shell geometry.

The positive values are of all stress resultant and stress couple quantities are shown on the shell element below.

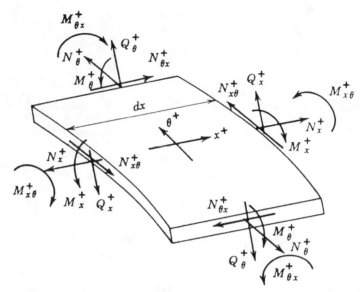

Figure 10.2. Positive directions of integrated stress quantities.

In shell theory, one uses all the assumptions used in the plate theory derivation of Chapter 2, such as:

$$\sigma_z = \varepsilon_r = \varepsilon_{zx} = \varepsilon_{z\theta} = 0$$

$$u_{\text{TOT}} = u_0(x, \theta) + \beta_x(x, \theta)z$$

$$v_{\text{TOT}} = v_0(x, \theta) + \beta_\theta(x, \theta)z \tag{10.1}$$

$$w = w(x, \theta).$$

In addition, there is one another condition known as Love's First Approximation, which is consistent with the neglect of transverse shear deformation. This is written as

$$h/R \ll 1. \tag{10.2}$$

To derive the governing equations for cylindrical shells properly, one should begin with the equations of elasticity in a curvilinear coordinate system, just as was done with the elasticity equations in a Cartesian coordinate system in Chapter 1 for the plate equation derivation. One can then proceed to develop the governing equations for a shell of general shape; specialize the equations to a shell of revolution, and finally specialize them to a shell of circular cylindrical geometry. To do this properly would require another two chapters, two weeks of lectures, and a knowledge of curvilinear coordinate systems; hence, it will not be done rigorously herein. Those interested should pursue a graduate course in shell theory. In what follows, please accept the governing equations as correct, and regard them as a starting point for the detailed treatment of solutions for circular cylindrical shells, and a vehicle to better understand the behavior of shells of all shapes.

The governing equations are as follows for a circular cylindrical shell subjected to in-plane and lateral distributed loads.

$$\frac{\partial N_x}{\partial x} + \frac{1}{R}\frac{\partial N_{x\theta}}{\partial \theta} + q_x = 0 \tag{10.3}$$

$$\frac{\partial N_{x\theta}}{\partial x} + \frac{1}{R}\frac{\partial N_\theta}{\partial \theta} + \frac{Q_\theta}{R} + q_\theta = 0 \tag{10.4}$$

$$\frac{\partial Q_x}{\partial x} + \frac{1}{R}\frac{\partial Q_\theta}{\partial \theta} - \frac{1}{R}N_\theta + p(x, \theta) \tag{10.5}$$

$$\frac{\partial M_x}{\partial x} + \frac{1}{R}\frac{\partial M_{x\theta}}{\partial \theta} - (Q_x - m_x) = 0 \tag{10.6}$$

$$\frac{\partial M_{x\theta}}{\partial x} + \frac{1}{R}\frac{\partial M_\theta}{\partial \theta} - (Q_\theta - m_\theta) = 0 \tag{10.7}$$

$$\beta_x + \frac{\partial w}{\partial x} = 0 \tag{10.8}$$

$$\beta_\theta + \frac{1}{R}\frac{\partial w}{\partial \theta} - \frac{v_0}{R} = 0 \tag{10.9}$$

$$N_x = K\left[\frac{\partial u_0}{\partial x} + \frac{v}{R}\frac{\partial v_0}{\partial \theta} + \frac{v}{R}w\right] \tag{10.10}$$

$$N_\theta = K\left[\frac{1}{R}\frac{\partial v_0}{\partial \theta} + \frac{w}{R} + v\frac{\partial u_0}{\partial x}\right] \tag{10.11}$$

$$N_{x\theta} = N_{\theta x} = (1 - v)\frac{K}{2}\left[\frac{1}{R}\frac{\partial u_0}{\partial \theta} + \frac{\partial v_0}{\partial x}\right] \tag{10.12}$$

$$M_x = D\left[\frac{\partial \beta_x}{\partial x} + \frac{v}{R}\frac{\partial \beta_\theta}{\partial \theta}\right] \tag{10.13}$$

$$M_\theta = D\left[\frac{1}{R}\frac{\partial \beta_\theta}{\partial \theta} + v\frac{\partial \beta_x}{\partial x}\right] \tag{10.14}$$

$$M_{x\theta} = M_{\theta x} = \frac{(1 - v)D}{2}\left[\frac{\partial \beta_\theta}{\partial x} + \frac{1}{R}\frac{\partial \beta_x}{\partial \theta}\right] \tag{10.15}$$

where

$$
\left\{
\begin{array}{c}
N_x \\
N_\theta \\
N_{x\theta} \\
Q_x \\
Q_\theta
\end{array}
\right\}
= \int_{-h/2}^{+h/2}
\left\{
\begin{array}{c}
\sigma_x \\
\sigma_\theta \\
\sigma_{x\theta} \\
\sigma_{xz} \\
\sigma_{\theta z}
\end{array}
\right\} dz ,
\qquad
\left\{
\begin{array}{c}
M_x \\
M_\theta \\
M_{x\theta}
\end{array}
\right\}
= \int_{-h/2}^{h/2}
\left\{
\begin{array}{c}
\sigma_x \\
\sigma_\theta \\
\sigma_{x\theta}
\end{array}
\right\} z \, dz
\tag{10.16}
$$

and

$$
q_x = \sigma_{zx}(h/2) - \sigma_{zx}(-h/2) = \tau_{1x} - \tau_{2x}
\tag{10.17}
$$

$$
q_\theta = \sigma_{z\theta}(h/2) - \sigma_{z\theta}(-h/2) = \tau_{1\theta} - \tau_{2\theta}
\tag{10.18}
$$

$$
m_x = \frac{h}{2}\left[\sigma_{zx}(h/2) + \sigma_{zx}(-h/2)\right] = \frac{h}{2}\left[\tau_{1x} + \tau_{2x}\right]
\tag{10.19}
$$

$$
m_\theta = \frac{h}{2}\left[\sigma_{z\theta}(h/2) + \sigma_{z\theta}(-h/2)\right] = \frac{h}{2}\left[\tau_{1\theta} + \tau_{2\theta}\right].
\tag{10.20}
$$

The above expressions for stress resultants, stress couples, and surface shear stresses are completely analogous to those for plates in Chapter 2.

Equations (10.3) through (10.5) are integrated equilibrium equations in the axial, circumferential, and radial direction, respectively. (10.6) and (10.7) are integrated moment equilibrium equations in the axial and circumferential directions. (10.8) and (10.9) provide values of the rotation β_x and β_θ terms of displacements, for the case of no transverse shear deformation, i.e., $\varepsilon_{xz} = \varepsilon_{\theta z} = 0$. (10.10) through (10.12) are integrated stress-displacement relations, while (10.13) through (10.15) are the cylindrical shell moment curvature relations (10.15) are the cylindrical shell moment curvature relations. In (10.16) the stress resultants and stress couples are defined, while (10.17) through (10.20) are surface shear stress quantities, analogous to those in plates discussed to Chapter 2.

The substitution of (10.8) and (10.9) into (10.13) through (10.15) give the stress couples in terms of the displacements u_0, v_0, and w. For simplicity, $R\theta = s$, the arc length in the circumferential direction:

$$
M_x = -D\left[\frac{\partial^2 w}{\partial x^2} + v\frac{\partial^2 w}{\partial s^2} - \frac{v}{R}\frac{\partial v_0}{\partial s}\right]
\tag{10.21}
$$

$$
M_x = -D\left[\frac{\partial^2 w}{\partial s^2} - \frac{1}{R}\frac{\partial v_0}{\partial s} + v\frac{\partial^2 w}{\partial x^2}\right]
\tag{10.22}
$$

$$
M_{xs} = M_{sx} = -\frac{(1-v)D}{2}D\left[2\frac{\partial^2 w}{\partial x\,\partial s} - \frac{1}{R}\frac{\partial v_0}{\partial x}\right].
\tag{10.23}
$$

For simplicity, in what follows we shall drop the terms involving the surface shears $q_x, q_s, m_x,$ and m_s. Substituting (10.21) through (10.23) into (10.6) and (10.7) provides the shear resultants in terms of the displacements.

$$Q_x = -D \left[\frac{\partial}{\partial x} (\nabla^2 w) - \frac{1 + v}{2R} \frac{\partial^2 v_0}{\partial x \, \partial s} \right] \tag{10.24}$$

$$Q_s = -D \left[\frac{\partial}{\partial s} (\nabla^2 w) - \frac{(1 - v)}{2R} \frac{\partial^2 v_0}{\partial x^2} - \frac{1}{R} \frac{\partial^2 v_0}{\partial s^2} \right] \tag{10.25}$$

where

$$\nabla^2(\ ) = \frac{\partial^2(\ )}{\partial x^2} + \frac{\partial^2(\ )}{\partial s^2} \ .$$

Substitution of (10.10) through (10.25) into the force equilibrium equations (10.3) through (10.5) provides three simultaneous equations in terms of u_0, v_0, and w, which are the final governing equations to solve.

$$\frac{\partial^2 u_0}{\partial x^2} + \frac{(1 - v)}{2} \frac{\partial^2 u_0}{\partial s^2} + \frac{1 + v}{2} \frac{\partial^2 v_0}{\partial x \, \partial s} + \frac{v}{R} \frac{\partial w}{\partial x} = 0 \tag{10.26}$$

$$\frac{(1 - v)}{2} \frac{\partial^2 v_0}{\partial x^2} + \frac{\partial^2 v_0}{\partial s^2} + \frac{1}{R} \frac{\partial w}{\partial s} + \frac{(1 + v)}{2} \frac{\partial^2 u_0}{\partial x \, \partial s} - k^2 R \frac{\partial}{\partial s} (\nabla^2 w) = 0 \tag{10.27}$$

$$\nabla^4 w - \frac{1}{R} \frac{\partial}{\partial s} (\nabla^2 v_0) + \frac{1}{k^2} \left[\frac{1}{R^3} \frac{\partial v_0}{\partial s} + \frac{1}{R^4} w + \frac{v}{R^3} \frac{\partial u_0}{\partial x} \right] = \frac{p(x, s)}{D} \tag{10.28}$$

where $k^2 = h^2/12R^2$.

In (10.26) through (10.28), Love's first approximation, given in (10.2), is utilized.

These governing equations form an eighth order system, which results in four boundary conditions on each edge. From application of variational principles (see Chapter 9), the resulting natural boundary conditions are:

For an edge that is $x = $ constant:

  Either u_0 prescribed or $N_x = 0$
  Either β_x prescribed or $M_x = 0$
  Either v_0 prescribed or $\overline{N}_{x\theta} = N_{x\theta} + (M_{x\theta}/R) = 0$
  Either w prescribed or $V_x = Q_x + (\partial M_{x\theta}/\partial s) = 0$.

For an edge that is $\theta = $ constant:

  Either v_0 prescribed or $N_\theta = 0$
  Either β_θ prescribed or $M_\theta = 0$
  Either u_0 prescribed or $\overline{N}_{\theta x} = N_{x\theta} = 0$
  Either w prescribed or $V_\theta = Q_\theta + (\partial M_{x\theta}/\partial x) = 0$.

The quantities $\overline{N}_x$ are 'effective' in-plane forces on the edges of the shell. V_x and V_θ are 'effective' transverse shear forces analogous to the Kirchoff boundary conditions discussed in Chapter 2 for a plate. These quantities are used because only four boundary conditions are obtained from the solution of the governing differential equations. Yet from physical reasoning on an x = constant edge which is free, the following boundary conditions must be satisfied: $N_x = N_{xs} = Q_x = M_x = M_{xs} = 0$. Since there are five, following the reasoning of Kirchoff the shear forces and twisting moments are combined. It should be noted that in shell theories which retain transverse shear deformation and transverse normal stress, the system is tenth order, and no approximations at the free boundary conditions is necessary.

For future reference, these effective boundary conditions are explicitly written in terms of the displacements.

$$\overline{N}_{xs} = \frac{(1-v)}{2}\left[K\left(\frac{\partial u_0}{\partial s} + \frac{\partial v_0}{\partial x}\right) - \frac{D}{R}\left(2\,\frac{\partial^2 w}{\partial x\,\partial s} - \frac{1}{R}\,\frac{\partial v_0}{\partial x}\right)\right] \tag{10.29}$$

$(x = \text{constant})$

$$\overline{N}_{xs} = \frac{1-v}{2}\,K\left[\frac{\partial u_0}{\partial s} + \frac{\partial v_0}{\partial x}\right] \tag{10.30}$$

$(\theta = \text{constant})$

$$V_x = -D\left[\frac{\partial^3 v_0}{\partial x^3} + (2-v)\,\frac{\partial^3 w}{\partial x\,\partial s^2} - \frac{1}{R}\,\frac{\partial^2 v_0}{\partial x\,\partial s}\right] \tag{10.31}$$

$$V_\theta = -D\left[\frac{\partial^3 w}{\partial s^3} + (2-v)\,\frac{\partial^3 w}{\partial x^2\,\partial s} - \frac{1-v}{R}\,\frac{\partial^2 v_0}{\partial x^2} - \frac{1}{R}\,\frac{\partial^2 v_0}{\partial s^2}\right]. \tag{10.32}$$

First putting all foregoing equations in terms of x and s ($= R\theta$), it should be noted that if in all the equations of this section the remaining values of R are set equal to infinity (let $R = \infty$), the equations reduce to those of a flat rectangular plate of coordinates x and s, of Chapter 2.

10.2. Circular Cylindrical Shells under Axially Symmetric Loads

When the loads on a circular cylindrical shell are axially symmetric, then from symmetry in the circumferential direction it is seen that $v_0 = 0$, and $\partial(\ )/\partial\theta = 0$.* It follows therefore from (10.7), (10.9), and (10.12), and (10.15) that $Q_\theta = \beta_\theta = N_{x\theta} = M_{x\theta} = 0$. The resulting governing differential equations are therefore (here surface shear stresses have been omitted purely for simplicity; but it is very simple to include them):

$$\frac{dN_x}{dx} = 0 \tag{10.33}$$

$$\frac{dQ_x}{dx} - \frac{N_\theta}{R} + p(x) = 0 \tag{10.34}$$

* There is one exception to this, that being a circular cylindrical shell subjected to an axially symmetric torsional load, $N_{x\theta}$, in which case $\partial(\ )/\partial\theta = 0$ but $v_0 \neq 0$.

$$\frac{dM_x}{dx} - Q_x = 0 \tag{10.35}$$

$$\beta_x + \frac{dw}{dx} = 0$$

$$N_x = K\left[\frac{du_0}{dx} + \frac{v}{R}w\right] \tag{10.36}$$

$$N_\theta = K\left[v\frac{du_0}{dx} + \frac{w}{R}\right] = vN_x + \frac{Ehw}{R} \tag{10.37}$$

$$M_x = -D\frac{d^2w}{dx^2} \tag{10.38}$$

$$M_\theta = -Dv\frac{d^2w}{dx^2} = vM_x \tag{10.39}$$

$$Q_x = -D\frac{d^3w}{dx^3} = V_x. \tag{10.40}$$

It is seen that from (10.33), N_x is a constant everywhere in the shell and is uniquely determined by the condition at the boundaries.

The governing equations in terms of displacements (10.26) through (10.28) reduce to two for the axially symmetric case, namely,

$$\frac{d^2u_0}{dx^2} + \frac{v}{R}\frac{dw}{dx} = 0 \tag{10.41}$$

$$\frac{d^4w}{dx^4} + \frac{1}{k^2R^4}w + \frac{v}{k^2R^3}\frac{du_0}{dx} = \frac{p(x)}{D}. \tag{10.42}$$

Solving (10.36) for du_0/dx, (10.42) can be written as

$$\frac{d^4w}{dx^4} + \frac{(1-v^2)}{k^2R^4}w = \frac{1}{D}\left[p(x) - v\frac{N_x}{R}\right]. \tag{10.43}$$

Substituting in the value for k^2, (10.43) can be written finally as

$$\frac{d^4w}{dx^4} + 4\varepsilon^4w = \frac{1}{D}\left[p(x) - v\frac{N_x}{R}\right] \tag{10.44}$$

where

$$\varepsilon^4 = \frac{3(1-v^2)}{h^2R^2}. \tag{10.45}$$

The form of the governing equation (10.44) is desirable since it is uncoupled from the other governing equation (10.41). N_x is a constant determined by boundary conditions. In fact it is seen lucidly that the presence of an axial in-plane force is that of an equivalent lateral pressure as far as the lateral displacement w is concerned.

It is also noted that the governing differential equation for the lateral deflection of a circular cylindrical shell has the same form as the governing differential equation for the lateral deflection of a beam on an elastic foundation; it would be identical if D were replaced by EI, and $4\varepsilon^4$ replaced by k, the foundation modulus. Thus, one may use the physical intuition, as well as the many known solutions for beams on an elastic foundation in considering these shells.

By standard methods, the roots of the fourth order equation (10.44) are determined to be $\pm \varepsilon(1 \pm i)$. Thus, the general solution can be written in the form

$$w(x) = Ae^{-\varepsilon x} \cos \varepsilon x + Be^{-\varepsilon x} \sin \varepsilon x + Ce^{\varepsilon x} \cos \varepsilon x$$
$$+ Ee^{\varepsilon x} \sin \varepsilon x + w_p(x), \tag{10.46}$$

where A, B, C, and E are constants of integration determined by the boundary conditions, and $w_p(x)$ is the particular integral. The in-plane displacement u can be determined by the first integral of (10.36), and is seen to be

$$u_0(x) = \frac{N_x x}{K} - \frac{v}{R} \int w \, dx + F \tag{10.47}$$

where F is a constant of integration.

It is seen that for the case of circular cylindrical shells under axially symmetric loads, there are six boundary conditions: four dealing with specifications of the lateral deflection slope, stress couple or shear resultant (w or its derivatives); the fifth is N_x, the in-plane stress resultant which is determined at the outset by external equilibrium; and the sixth (F) is determined by the specification of the in-plane displacement at some axial location.

Before proceeding with some solutions, a sketch of the shell showing the positive directions of displacements, and loads is in order, as presented in Figure 10.3.

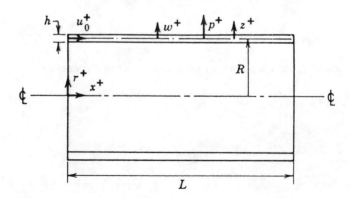

Figure 10.3. Circular cylindrical shell geometry, displacements, and coordinates.

The positive directions of stress resultants and couples is given in Figure 10.4.

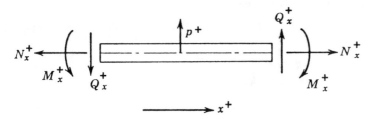

Figure 10.4. Circular cylindrical shell stress resultants and couples.

10.3. Edge Load Solutions

In the following, only the solutions for the lateral deflections are explicitly determined. The in-plane displacement u_0 can be subsequently determined easily from (10.47).

10.3.1. A Semi-Infinite Shell $(0 \leqslant x \leqslant \infty)$ Subjected to an Edge Moment $M_x = M_0$ at $x = 0$

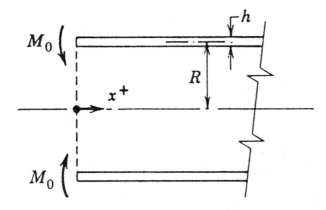

Figure 10.5. Circular cylindrical shell subjected to an edge moment at one end.

Since in this case $p(x) = N_x = 0$, only the homogeneous portion of the general solution (10.46) is needed. The boundary conditions at $x = 0$ are

$$M_x(0) = M_0 = -D \frac{d^2 w(0)}{dx^2} \tag{10.48}$$

$$Q_x(0) = 0 = -D \frac{d^3 w(0)}{dx^3} . \tag{10.49}$$

Since we are dealing with small displacements, using linear theory, it is seen that for w to remain finite as $x \to \infty$, it is required that $C = E = 0$. Hence,

$$w(x) = Ae^{-\varepsilon x} \cos \varepsilon x + Be^{-\varepsilon x} \sin \varepsilon x . \tag{10.50}$$

Substituting (10.50) into (10.48) and (10.49), the following are obtained.

$$w''(0) = -2\varepsilon^2 B = -\frac{M_0}{D}$$

$$w'''(0) = 2\varepsilon^3(A + B) = 0$$

where primes denote differentiation with respect to x. Thus, $B = -A = M_0/2\varepsilon^2 D$, and the solution is

$$w(x) = \frac{M_0}{2\varepsilon^2 D} e^{-\varepsilon x} (\sin \varepsilon x - \cos \varepsilon x) . \tag{10.51}$$

Of course, knowing $w(x)$ one can obtain M_x, Q_x, u_0 everywhere.

10.3.2. A Semi-Infinite Shell $(0 \leqslant x \leqslant \infty)$ Subjected to an Edge Shear $Q_x = Q_0$ at $x = 0$

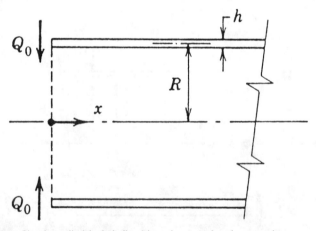

Figure 10.6. Circular cylindrical shell subjected to an edge shear resultant at one end.

Here the boundary conditions are

$$M_x(0) = 0 = -Dw'' \tag{10.52}$$

$$Q_x(0) = Q_0 - Dw'' \tag{10.53}$$

Again it is required that $C = E = 0$, and the solution is given by (10.50). Substituting (10.50) into (10.52) and (10.53) results in

$$2\varepsilon^2 B = 0$$

$$2\varepsilon^3(A + B) = -\frac{Q_0}{D} .$$

The solution is therefore

$$w(x) = -\frac{Q_0}{2\varepsilon^3 D} e^{-\varepsilon x} \cos \varepsilon x . \tag{10.54}$$

10.3.3. Edge Load Solutions as the Homogeneous Solution

It is seen that the solutions of (10.51) and (10.54) exhibit the same form: a constant times an oscillating (trigonometric) factor and a factor which exhibits an exponential decay away from the edge of the shell. This decay in the lateral deflection due to a edge stress couple or an edge transverse shear resultant is one of the characteristics of shells in general, and is one of the most important features of shell behavior. Since the slope, bending moment, and shear resultant away from the edge are all proportional to the derivatives of the lateral deflection, each of these also decays away from the edge where the edge load is acting. This characteristic is called the 'bending boundary layer', and it is seen that bending and shear stresses due to the edge load occur only in this bending boundary layer.

Now the lateral deflection, slope, bending moment, and transverse shear resultant all decay as $e^{-\varepsilon x}$ where $\varepsilon = [3(1 - v^2)]^{1/4}/\sqrt{Rh}\}$. Hence, $x/\sqrt{Rh}$ is a fundamentally important parameter with regard to shell behavior. It is seen that for $x/\sqrt{Rh} \geqslant 4$, requires that $\varepsilon x \geqslant 5.15$ when $v = 0.3$ for example. This then means that $e^{-\varepsilon x} \leqslant 0.006$, and therefore the lateral deflection, slope, moment, and shear for $x/\sqrt{Rh} \geqslant 4$ are negligibly small. Therefore the length L_B of the bending boundary layer is taken to be*

$$L_B = 4Rh \quad 4\sqrt{Rh} * \tag{10.55}$$

This suggests a very useful solution technique for shell problems. Consider a finite length shell subjected to some axially symmetric loading and some set of stated boundary conditions. Instead of satisfying the boundary conditions directly through obtaining values of A, B, C, and E in (10.46), determine values of unknown edge loads, used as a form of the homogeneous solution, to satisfy the stated boundary conditions. The advantage of this method is that the effects of the particular boundary conditions on w and its derivatives become negligible for a distance L_B away from the edge. Further away from the edge only the particular solution will contribute to the lateral deflection, slope, bending moments, and transverse shear. Hence, if the length of the shell L is $L \geqslant L_B$, the boundary conditions involving w at one end of the shell are uncoupled from those at the other end. Mathematically, this means that for a shell of $L \geqslant L_B$, instead of solving a 4 × 4 matrix** to obtain the boundary conditions, one solves two 2 × 2 matrices.

---

* It can be shown that this can be generalized to any shell of revolution under axially symmetric edge load by stating that $L_B = 4\sqrt{R_0 h}$, where R_0 is the circumferential radius of curvature at the edge.

** Actually a 6 × 6 matrix for a general shell of revolution under axially symmetric load, but for cylindrical shells, N_x is found by a trivial force balance and the boundary constant F in the $u_0(x)$ equation is always found subsequent to finding the four boundary constants involving w and its derivatives.

Using the edge load form of the solution, the homogeneous solution for a shell of length L is written as

$$w(x) = \frac{M_0}{2\varepsilon^2 D} e^{-\varepsilon x} (\sin \varepsilon x - \cos \varepsilon x) - \frac{Q_0}{2\varepsilon^3 D} e^{-\varepsilon x} \cos \varepsilon x$$

$$+ \frac{M_L}{2\varepsilon^2 D} e^{-\varepsilon(L-x)} [\sin \varepsilon(L-x) - \cos \varepsilon(L-x)]$$

$$+ \frac{Q_L}{2\varepsilon^3 D} e^{-\varepsilon(L-x)} \cos \varepsilon(L-x)$$

where the edge loads are considered positive as shown in Figure 10.7.

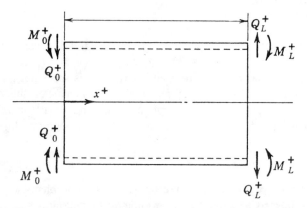

Figure 10.7. Positive directions for edge loads on a circular cylindrical shell.

It should be noted from (10.47) that only some of the terms in the $u_0(x)$ equation decay away from the edges, namely the second term as written. The others involve one that increases monotonically in x, the other is a constant. In any case outside the bending boundary layer at each edge, the expression for u_0 is simplified.

Even when the length of the shell is so short that $L = L_B$, in which case there is no separation of the boundary conditions, there is no practical advantage in using the form of the homogeneous solution given by (10.46) compared with the form of (10.55).

10.4. A General Solution for Cylindrical Shells under Axially Symmetric Loads

For reference in solving problems of this kind, all equations needed are catalogued below. (See Figure 10.7.) It is restricted to cases where $d^4 p(x)/dx^4 = 0$. Since this would almost always be true from a practical point of view, i.e., very few loads would have a non-zero fourth derivative, the solutions presented below are fairly general.

$$w(x) = \frac{M_0}{2\varepsilon^2 D} e^{-\varepsilon x} (\sin \varepsilon x - \cos \varepsilon x) - \frac{Q_0}{2\varepsilon^3 D} e^{-\varepsilon x} \cos \varepsilon x$$

$$+ \frac{M_L}{2\varepsilon^2 D} e^{-\varepsilon(L-x)} [\sin \varepsilon(L-x) - \cos \varepsilon(L-x)]$$

$$+ \frac{Q_L}{2\varepsilon^3 D} e^{-\varepsilon(L-x)} \cos \varepsilon(L-x) + \frac{1}{4\varepsilon^4 D} \left[p(x) - \frac{\nu N_x}{R} \right] \qquad (10.56)$$

where

$$\varepsilon = [3(1-v^2)]^{1/4}/\sqrt{Rh}$$

$$w'(x) = \frac{dw}{dx} = \frac{M_0}{\varepsilon D} e^{-\varepsilon x} \cos \varepsilon x + \frac{Q_0}{2\varepsilon^2 D} e^{-\varepsilon x} (\sin \varepsilon x + \cos \varepsilon x)$$

$$- \frac{M_L}{\varepsilon D} e^{-\varepsilon(L-x)} \cos \varepsilon(L-x)$$

$$+ \frac{Q_L}{2\varepsilon^2 D} e^{-\varepsilon(L-x)} [\sin \varepsilon(L-x) + \cos \varepsilon(L-x)] + \frac{1}{4\varepsilon^4 D} p'(x) \quad (10.57)$$

$$M_x(x) = -Dw''(x) = M_0 e^{-\varepsilon x} (\sin \varepsilon x + \cos \varepsilon x) + \frac{Q_0}{\varepsilon} e^{-\varepsilon x} \sin \varepsilon x$$

$$+ M_L e^{-\varepsilon(L-x)} [\sin \varepsilon(L-x) + \cos \varepsilon(L-x)]$$

$$- \frac{Q_L}{\varepsilon} e^{-\varepsilon(L-x)} \sin \varepsilon(L-x) - \frac{1}{4\varepsilon^4} p''(x) \qquad (10.58)$$

$$Q_x = -Dw'''(x) = -2M_0\varepsilon e^{-\varepsilon x} \sin \varepsilon x$$

$$+ Q_0 e^{-\varepsilon x} (\cos \varepsilon x - \sin \varepsilon x) + 2M_L\varepsilon e^{-\varepsilon(L-x)} \sin \varepsilon(L-x)$$

$$- Q_L e^{-\varepsilon(L-x)} [-\cos \varepsilon(L-x) + \sin \varepsilon(L-x)] - \frac{1}{4\varepsilon^4} p'''(x) \qquad (10.59)$$

$$u_0(x) = \left[\frac{1}{K} + \frac{v^2}{4R^2\varepsilon^4 D} \right] N_x x - \frac{v}{R} \left\{ -\frac{M_0}{2\varepsilon^3 D} e^{-\varepsilon x} \sin \varepsilon x \right.$$

$$+ \frac{Q_0}{4\varepsilon^4 D} e^{-\varepsilon x} (\cos \varepsilon x - \sin \varepsilon x) + \frac{M_L}{2\varepsilon^3 D} e^{-\varepsilon(L-x)} \sin \varepsilon(L-x)$$

$$+ \frac{Q_L}{4\varepsilon^4 D} e^{-\varepsilon(L-x)} [\cos \varepsilon(L-x) - \sin \varepsilon(L-x)] \left. \right\} - \frac{v}{4R\varepsilon^4 D} \int p(x)\,dx + F$$

$$(10.60)$$

$$N_x = \text{constant} \tag{10.61}$$

$$N_\theta(x) = vN_x + \frac{Ehw(x)}{R} \tag{10.62}$$

$$M_\theta(x) = vM_x(x) \tag{10.63}$$

$$\sigma_x = \frac{N_x}{h} + \frac{M_x z}{h^3/12} \tag{10.64}$$

$$\sigma_\theta = \frac{N_\theta}{h} + \frac{M_\theta z}{h^3/12} = v\sigma_x + \frac{Ew(x)}{R} \tag{10.65}$$

$$\sigma_{xz} = \frac{3Q_x}{2h}\left[1 - \left(\frac{z}{h/2}\right)^2\right]. \tag{10.66}$$

The edge load solutions for a conical shell are provided by Vinson (Reference 10.1), and the edge load solutions for hemispherical shells are given by Reference 10.2. Since these are often connected to cylindrical shells in many practical structures, their use is frequent.

Also problems involving cylindrical shells of composite materials are treated by Vinson and Sierakowski in Reference 10.3.

10.5. Sample Solutions

10.5.1. Effects of Simple and Clamped Supports

Consider the circular cylindrical shell shown in Figure 10.8. The end of the shell $x = 0$ is simply supported, the end $x = L$ is clamped. The plate ends of the shell are assumed rigid.

The internal pressure is p_0, and $v = 0.3$. The following information is desired:

a. What is the magnitude and exact location of the maximum stress that occurs in the bending boundary layer at the simply supported end?
b. The same information near the clamped end?
c. What is the magnitude of the maximum stress occurring outside the bending boundary layers, say at $x = L/2$?
d. If the maximum principal stress failure theory is used, would the shell be structurally sound if in the design the thickness had been determined by membrane shell theory? (Membrane shell theory neglects all bending effects, see Section 10.10),

From external axial force equilibrium, $N_x = p_0 R/2$. The boundary conditions at $x = 0$ are $w(0) = 0$, and $M(0) = 0$. From (10.56) and (10.58), the boundary conditions are:

$$w(0) = -\frac{M_0}{2\varepsilon^2 D} - \frac{Q_0}{2\varepsilon^3 D} + \frac{1}{4\varepsilon^4 D}\, p_0(1 - v/2) \tag{10.67}$$

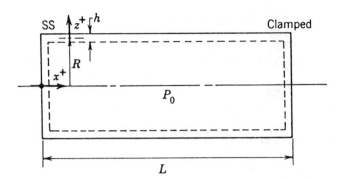

Figure 10.8. Example problem of shell of constant internal pressure with one end simply supported and the other end clamped.

$$M_x(0) = M_0 = 0 \tag{10.68}$$

Hence,

$$Q_0 = \frac{p_0}{2\varepsilon}(1 - v/2). \tag{10.69}$$

To determine the location of the maximum value of σ_x, the location of the maximum value of $M(x)$ is required since N_x is a constant. Since M_x is zero at $x = 0$, and again it is zero for $x \geqslant L_B$, extremum values lie somewhere in the bending boundary layer. They occur at values of x where $dM_x/dx = 0$. Since $dM_x/dx = Q_x$, therefore the extremes occur at $Q_x = 0$. From (10.68), the fact that $p_0 = $ constant, and since the shell is long, the effects of M and Q are negligible, (10.59) becomes $Q_x = (p_0/2)$ $(1 - v/2)e^{-\varepsilon x}(\cos \varepsilon x - \sin \varepsilon x) = 0$. For this condition to exist, $\cos \varepsilon x = \sin \varepsilon x$, or $\varepsilon x = \pi/4, 3\pi/4, 5\pi/4$, etc. Since there is an exponential decay with increasing x, the largest extreme value, the maximum occurs at $\varepsilon x = \pi/4$. Hence,

$$M_{x_{max}} = M_x\left(\frac{\pi}{4\varepsilon}\right) = p_0(1 - v/2)\frac{\sqrt{2}}{4\varepsilon^2}e^{-\pi/4}. \tag{10.70}$$

From (10.64)

$$\sigma_{x_{max}} = \frac{N_x}{h} \pm \frac{6M_{x_{max}}}{h^2}$$

$$= \frac{p_0 R}{2h} \pm 0.498\frac{p_0 R}{h}, \quad \text{for} \quad v = 0.3$$

$$\sigma_{x_{max}} = \sigma_x\left(\frac{\pi}{4\varepsilon}, +h/2\right) = 0.998\frac{p_0 R}{h}. \tag{10.71}$$

From (10.68), (10.69), (10.56), and (10.58),

$$w(x) = \frac{p_0(1 - v/2)}{4\varepsilon^4 D} [1 - e^{-\varepsilon x} \cos \varepsilon x] \quad \text{for} \quad x \leqslant L_B \tag{10.72}$$

$$M(x) = \frac{p_0(1 - v/2)}{2\varepsilon^2} e^{-\varepsilon x} \sin \varepsilon x \quad \text{for} \quad x \leqslant L_B \tag{10.73}$$

Therefore from (10.65), (10.72), and (10.73).

$$\sigma_\theta = \frac{p_0 R}{h} \pm \frac{3v(1 - v/2)R p_0 e^{-\varepsilon x} \sin \varepsilon x}{h[3(1 - v^2)]^{1/2}}$$

$$- \frac{p_0(1 - v/2)R e^{-\varepsilon x} \cos \varepsilon x}{h} \quad \text{for} \quad z = \pm h/2 . \tag{10.74}$$

For $v = 0.3$, this reduces to

$$\sigma_\theta = \frac{p_0 R}{h} \{1 \pm 0.464 e^{-\varepsilon x} \sin \varepsilon x - 0.85 e^{-\varepsilon x} \cos \varepsilon x\} \quad \text{for} \quad x \leqslant L_B, z = \pm h/2 .$$

Extremum values occur for the condition $\partial \sigma_\theta / \partial x = 0$, which results in the requirement that

$$\pm 0.464 (- \sin \varepsilon x + \cos \varepsilon x) + 0.85 (\sin \varepsilon x + \cos \varepsilon x) = 0 . \tag{10.75}$$

The + requirement occurs that $\varepsilon x = 1.875$; the negative when $\varepsilon x = 2.845$. Of these two, σ_θ is maximum for the former. The maximum value of σ_θ in the range $x \leqslant L_B$, is

$$\sigma_{\theta_{max}} = \sigma_\theta \left(\frac{1,875}{\varepsilon} , +h/2 \right) = 1.1072 \frac{p_0 R}{h} \quad \text{for} \quad v = 0.3 . \tag{10.76}$$

At the clamped end, the boundary conditions are $w(L) = w'(L) = 0$. From (10.56), (10.57), and making use of the fact that the shell is long $(L > 4\sqrt{Rh})$

$$w(L) = 0 = -\frac{M_L}{2\varepsilon^2 D} + \frac{Q_L}{2\varepsilon^3 D} + \frac{p_0}{4\varepsilon^4 D} (1 - v/2) = 0$$

$$w'(L) = 0 = -\frac{M_L}{\varepsilon D} + \frac{Q_L}{2\varepsilon^2 D} = 0 .$$

Hence,

$$M_L = -\frac{p_0(1 - v/2)}{2\varepsilon^2} \tag{10.77}$$

$$Q_L = -\frac{p_0(1 - v/2)}{\varepsilon} \tag{10.78}$$

It can be shown and is physically obvious that at the clamped end, $M_{x_{max}} = M(L) = M_L$.
Hence, at $x = L$,

$$\sigma_x = \frac{N_x}{h} + \frac{M_L z}{h^3/12} = \frac{p_0 R}{2h} - \frac{p_0(1 - v/2)z}{2\varepsilon^2(h^3/12)}$$

(10.79)

$$\sigma_{x_{max}} = \sigma_x(L, -h/2) = 2.04 \frac{p_0 R}{h} \quad \text{for} \quad v = 0.3 .$$

However, $\sigma_{\theta_{max}}$ occurs away from the end of the shell. Analogous to the procedures used at the other end, it is found that

$$\sigma_{\theta_{max}} = \sigma_\theta \left(L - \frac{2,65}{\varepsilon} , +h/2 \right) = 1.069 \frac{p_0 R}{h} .$$

(10.80)

At $x = L/2$, which is outside either bending boundary layer, it is seen from (10.58) that $M_x = 0$, hence,

$$\sigma_x = \frac{N_x}{h} = \frac{p_0 R}{2h}$$

(10.81)

which is the membrane solution.
Likewise,

$$\sigma_\theta = v\sigma_x + \frac{Ew}{R} = \frac{p_0 R}{h} .$$

(10.82)

This is also the membrane solution.

It is seen that the maximum principal stress occurring in the boundary layer at the simply supported end is, from (10.76), 1.1072 $p_0 R/h$; correspondingly, at the clamped end it is 2.04 $p_0 R/h$ (from 10.79). The maximum stress predicted by membrane theory is 1.0 $p_0 R/h$. Hence, stresses greater than membrane stresses occur in both boundary layers; 10% higher in the simply supported area, and 104% higher in the clamped edge. Thus, this shell if designed on a basis of membrane shell theory would be woefully inadequate.

In the above, we have determined the location of the maximum axial stress and the maximum circumferential stress at each end of the shell. It should be remembered that everywhere there is a biaxial stress state. Hence, when dealing with a material which follows yield or fracture criteria such as a maximum distortion energy criteria, maximum shear stress, or one of several others, then that criteria must be used to find the location of the maximum value of the equivalent uniaxial stress state. Quite often a very simple digital computer routine can be employed to do the arithmetic, using the analytical solution, to determine the location and magnitude of the maximum stress state.

10.5.2. Supports by Elastic Rings

Consider the shell in the previous subsection, however, each shell end is supported by an elastic ring which has negligible torsional rigidity. In that case $M(0) = M(L) = M_0 = M_L = 0$.
From (10.56)

$$w(x) = -\frac{Q_0}{2\varepsilon^3 D} e^{-\varepsilon x} \cos\varepsilon x + \frac{Q_L}{2\varepsilon^3 D} e^{-\varepsilon(L-x)} \cos\varepsilon(L-x) + \frac{p_0(1-v/2)}{4\varepsilon^4 D} .$$

(10.83)

Force equilibrium on a segment of each ring requires the following (where all subscripts r refer to the ring):

$$\sigma_{0r} = -\frac{RQ_0}{A_r} = +\frac{RQ_L}{A_r}$$

(10.84)

where A_r = ring cross-sectional area.
From ring theory,

$$\varepsilon_{0r} = \frac{w_r}{R} = \frac{\sigma_{0r}}{E_r} = -\frac{RQ_0}{A_r E_r} = \frac{RQ_L}{A_r E_r} .$$

(10.85)

For the ring to support the shell, it is required that

$$w(0) = w_R \quad \text{and} \quad w(L) = w_R .$$

(10.86)

Hence, the remaining boundary conditions are:

$$-\frac{Q_0 R^2}{A_r E_r} = -\frac{Q_0}{2\varepsilon^3 D} + \frac{p_0(1-v/2)}{4\varepsilon^4 D}$$

(10.87)

$$\frac{Q_L R^2}{A_r E_r} = \frac{Q_L}{2\varepsilon^3 D} + \frac{p_0(1-v/2)}{4\varepsilon^4 D} .$$

(10.88)

Solving for Q_0 and Q_L, it is seen that

$$w(x) = \frac{p_0(1-v/2)}{4\varepsilon^4 D} \left[1 - \frac{1}{1 + \dfrac{2\varepsilon^3 D R^2}{E_r A_r}} \right] [e^{-\varepsilon x} \cos\varepsilon x + e^{-\varepsilon(L-x)} \cos\varepsilon(L-x)]$$

(10.89)

From this solution, all other quantities are obtained from the equations of Section 10.4.

Note that if the rings are infinitely stiff ($E_R = \infty$), the solution reduces to that of the shell on simple supports. Also note that outside the bending boundary layers at each end, only the particular solution remains, which is the membrane solution.

Also, if the ring did have significant torsional stiffness, then the stress couple of the shell at each end could be related to the torsional stiffness of the ring, analogous to the above.

Also, in a cylindrical shell, if the ring spacing is larger than the bending boundary layer, i.e., $> 4\sqrt{Rh}$, each ring-shell area can be analyzed separately. If the spacing is smaller, then interaction will occur, but the procedure is straightforward.

10.5.3. An Infinite Shell $(-\infty \leqslant x \leqslant +\infty)$ Subjected to a Radial Line Road H at $x = 0$

In this case $p(x) = 0$, $N_x = 0$, and only the homogeneous solution is used. Since the shell and loading are symmetric with respect to $x = 0$, we need only consider the half of the shell $0 \leqslant x \leqslant \infty$. Also, from symmetry, it is easily deduced that $Q_0 = -H/2$. M_0, the edge moment, is then used to enforce the boundary condition the slope at $x = 0$ be zero.

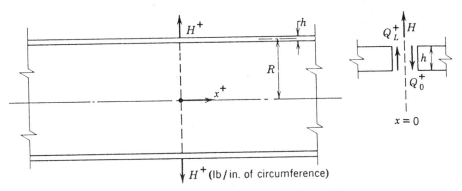

Figure 10.9. Cylindrical shell with radial line load H.

Thus, combining (10.56) and (10.57) we obtain

$$w(x) = \frac{M_0}{2\varepsilon^2 D} e^{-\varepsilon x} (\sin \varepsilon x - \cos \varepsilon x) - \frac{Q_0}{2\varepsilon^3 D} e^{-\varepsilon x} \cos \varepsilon x \qquad (10.90)$$

$$w'(x) = \frac{M_0}{\varepsilon D} e^{-\varepsilon x} \cos \varepsilon x + \frac{Q_0}{2\varepsilon^2 D} e^{-\varepsilon x} (\cos \varepsilon x + \sin \varepsilon x) \qquad (10.91)$$

and at $x = 0$

$$w'(0) = 0 = \frac{M_0}{\varepsilon D} + \frac{Q_0}{2\varepsilon^2 D} . \qquad (10.92)$$

Since $Q_0 = -H/2$, then $M_0 = H/4\varepsilon$.

The solution is then written as

$$w(x) = \frac{H}{8\varepsilon^3 D} e^{-\varepsilon x} (\sin \varepsilon x + \cos \varepsilon x) \qquad (10.93)$$

As previously mentioned, the lateral deflection is symmetric with respect to $x = 0$.

Again, the decay $e^{-\varepsilon x}$ is present, just as in the case of edge loads. This result can be generalized. In a region of the bending boundary lawyer, a dimension $4\sqrt{Rh}$ from any edge, structural discontinuity or load discontinuity, there will exist bending stresses superimposed on the membrane stresses; outside of this region only membrane stresses (and displacements) will exist. This is true for any shell of revolution if R, the circumferential radius of curvature, is used in determining the length of the bending boundary layer.

10.6. Circular Cylindrical Shells under Asymmetric Loads

The full set of equations for circular cylindrical shells under an asymmetric load is given by Equations (10.3) through (10.15). The applied loads are $p(x, \theta)$ and

$$
\begin{aligned}
\tau_{1x} &= \sigma_{rx}(+h/2) \\
\tau_{2x} &= \sigma_{rx}(-h/2) \\
\tau_{1\theta} &= \sigma_{r\theta}(+h/2) \\
\tau_{2\theta} &= \sigma_{r\theta}(-h/2)
\end{aligned}
\tag{10.94}
$$

The quantities defined in (10.94) are *known* applied surface shear loadings for a single layer shell, or the outer surface of a multilayer shell. However, on the inner surfaces of a multilayer shell such quantities are *unknown* and are included among the dependent variables for which solutions must be found.

In most asymmetric loads, it is convenient to expand the applied loads into Fourier Series as shown below:

$$
f(x, \theta) = \sum_{n=0}^{\infty} f_n(x) \begin{Bmatrix} \cos n\theta \\ \sin n\theta \end{Bmatrix}
\tag{10.95}
$$

where $f(x, \theta)$ are $p(x, \theta)$, τ_{1x}, τ_{2x}, $\tau_{1\theta}$, and $\tau_{2\theta}$, which could be continuous or discontinuous functions.

Likewise, all dependent variables are expanded into Fourier Series:

$$
g(x, \theta) = \sum_{n=0}^{\infty} g_n(x) \begin{Bmatrix} \cos n\theta \\ \sin n\theta \end{Bmatrix}
\tag{10.96}
$$

when $g(x, \theta)$ can be w, u_0, v_0, M_x, M, M_x, N_x, N, N_x, N_x, Q_x, Q_θ, β_x, and β_θ.

Substituting (10.95) and (10.96) into (10.33) through (10.15) results in a complete set of ordinary differential equations in x. Obviously, proper substitution will result in three simultaneous differential equations, in U_n, V_n, and W_n with constant coefficients, analogous to (10.26) through (10.28).

These equations are straightforward in solution. An eighth order algebraic equation is obtained from which the roots are found. Because of the complexity of the constant coefficients, the roots are usually obtained after substitution of numbers for geometry and material properties, rather than in algebraic generality. When obtaining the numerical roots, care must be exercised in retaining sufficient accuracy (significant figures) that subsequent differentiation of the displacements to find the stresses will not introduce significant errors. Also, for any other geometry, material, or load the roots must again be obtained for each solution if numerical rather than geometrically general coefficients are employed.

An alternative to the foregoing is to utilize a simplified shell theory in which the roots are obtained in algebraically general form.

One such simplified shell theory is known as shallow shell theory.

10.7. Shallow Shell Theory (Donnell's Equations)

Donnell formulated his theory in 1933 in his study of elastic stability of cylindrical shells. Since that time the approximations have been used in stress and vibration analysis extensively. The shallow shell equations are simpler than the full set, and can be decoupled. Further, for cylindrical shells the roots can be obtained with algebraic generality.

The shallow shell equations are accurate for thin shells except near to a very localized loading.

Pohle and Hoff have systematically compared Donnell shallow shell theory results with those of the more complicated shell theory of Flügge. Results for $n = 1$ resulted in the greatest difference, and it was only a few percent. For $n \gg 1$ Donnell theory solutions accurately match those of Flügge.

Donnell-type equations, including transverse shear deformation, which incorporate the ε/R_0 terms, have been found by R. M. Cooper to be accurate for thick shells subjected to line loads. For $R/h = 10$, the Cooper solution agrees very well with a three-dimensional elasticity solution, according to J. Greenspon.

To obtain the Donnell equations from the full set, the following two assumptions are made (*Note*: the following assumptions are made in transforming shell equations of any geometry to shallow shell equations for that geometry, and are not limited to cylindrical shells):

1. Use the moment-curvature relations for plates (i.e., neglect initial curvature).
2. Neglect the effect of transverse shear force, Q, on the balance of forces in the circumferential direction.

Employing these two assumptions, the general set of Equations (10.3) through (10.15) become (neglecting surface shear forces):

$$\frac{\partial N_x}{\partial x} + \frac{1}{R} \frac{\partial N_{x\theta}}{\partial \theta} = 0 \tag{10.97}$$

$$\frac{\partial N_{x0}}{\partial x} + \frac{1}{R} \frac{\partial N_\theta}{\partial \theta} = 0* \tag{10.98}$$

$$\frac{\partial Q_x}{\partial x} + \frac{1}{R} \frac{\partial Q_\theta}{\partial \theta} - \frac{1}{R} N_\theta + p(x, \theta) = 0 \tag{10.99}$$

$$\frac{\partial M_x}{\partial x} + \frac{1}{R} \frac{\partial M_{x\theta}}{\partial \theta} - Q_x = 0 \tag{10.100}$$

$$\frac{\partial M_{x\theta}}{\partial x} + \frac{1}{R} \frac{\partial M_\theta}{\partial \theta} - Q_\theta = 0 \tag{10.101}$$

$$N_x = K \left[\frac{\partial u_0}{\partial x} + \frac{v}{R} \frac{\partial v_0}{\partial \theta} + \frac{v}{R} w \right] \tag{10.102}$$

$$N_\theta = K \left[\frac{1}{K} \frac{\partial v_0}{\partial \theta} + v \frac{\partial u_0}{\partial x} + \frac{w}{R} \right] \tag{10.103}$$

$$N_{x\theta} = N_{\theta x} = \frac{(1-v)K}{2} \left[\frac{1}{R} \frac{\partial u_0}{\partial \theta} + \frac{\partial v_0}{\partial x} \right] \tag{10.104}$$

$$M_x = -D \left[\frac{\partial^2 w}{\partial x^2} + \frac{v}{R^2} \frac{\partial^2 w}{\partial \theta^2} \right]^* \tag{10.105}$$

$$M_\theta = -D \left[\frac{1}{R^2} + \frac{\partial^2 w}{\partial \theta^2} + v \frac{\partial^2 w}{\partial x^2} \right]^* \tag{10.106}$$

$$M_{x\theta} = M_{\theta x} = -\frac{(1-v)D}{R} \frac{\partial^2 w}{\partial x \, \partial \theta}^* \tag{10.107}$$

$$Q_x = -D \frac{\partial}{\partial x} (\nabla^2 w)^* \tag{10.108}$$

$$Q_\theta = -\frac{D}{R} \frac{\partial}{\partial \theta} (\nabla^2 w)^* . \tag{10.109}$$

The effective shear, associated with free edges becomes:

$$V_x = -D \left[\frac{\partial^3 w}{\partial x^3} + \frac{(2-v)}{R} \frac{\partial^3 w}{\partial x \, \partial \theta^2} \right]^* \tag{10.110}$$

$$V_\theta = -D \left[\frac{1}{R} \frac{\partial^3 w}{\partial \theta^3} + \frac{2-v}{R} \frac{\partial^3 w}{\partial x^2 \, \partial \theta} \right]^* \tag{10.111}$$

In all the above, the asterisk denotes equations which differ from those of the complete unsimplified set.

Proceeding as before in Section 10.2, three simultaneous force equilibrium equations are obtained in terms of the two mid-surface displacements, u_0 and v_0, and the lateral displacement w.

$$\frac{\partial^2 u_0}{\partial x^2} + \frac{(1-v)}{2} \frac{\partial^2 u_0}{\partial s^2} + \frac{(1+v)}{2} \frac{\partial^2 v_0}{\partial x \, \partial s} + \frac{v}{R} \frac{\partial w}{\partial x} = 0 \tag{10.112}$$

$$\frac{(1 - v)}{2} \frac{\partial^2 v_0}{\partial x^2} + \frac{\partial^2 v_0}{\partial s^2} + \frac{1 + v}{2} \frac{\partial^2 u_0}{\partial x \partial s} + \frac{1}{R} \frac{\partial w}{\partial s} = 0 \qquad (10.113)$$

$$\nabla^4 w + \frac{12}{h^2 R} \left[\frac{\partial v_0}{\partial s} + \frac{w}{R} + v \frac{\partial u_0}{\partial x} \right] = \frac{p(x, s)}{D} . \qquad (10.114)$$

A comparison of these with (10.26) through (10.28) provides a clear picture of what simplifications result from the shallow shell assumptions. Manipulation of the three equations above results in the Donnell Equations shown below in their familiar form.

$$\nabla^4 u_0 = -\frac{1}{R} \frac{\partial}{\partial x} \left[v \frac{\partial^2 w}{\partial x^2} - \frac{\partial^2 w}{\partial s^2} \right] \qquad (10.115)$$

$$\nabla^4 v_0 = -\frac{1}{R} \frac{\partial}{\partial s} \left[(2 + v) \frac{\partial^2 w}{\partial x^2} + \frac{\partial^2 w}{\partial s^2} \right] \qquad (10.116)$$

$$\nabla^8 w + 4\varepsilon^4 \frac{\partial^4 w}{\partial x^4} = \frac{1}{D} \nabla^4 p(x, s) \qquad (10.117)$$

It is noted the last equation is uncoupled from the first two. Upon its solution, the first two are subsequently solved by proper substitution of w into the right hand sides of each.

If the circular cylindrical shell is complete in the circumferential direction $0 \leqslant \theta \leqslant 2\pi$, one approach to the solution is to reduce the above to three ordinary differential equations in x, by using the form of the displacements u_0, v_0, and w given in (10.96), and the form of the lateral load $p(x, \theta)$ given in (10.95). Proceeding in this manner, the governing equations for the homogeneous solution are given as follows:

$$U_n^{IV} - \frac{2n^2}{R^2} U_n'' + \frac{n^4}{R^4} U_n = \frac{1}{R} \left[-\frac{n^2}{R^2} W_n' - v W_n''' \right] \qquad (10.118)$$

$$V_n^{IV} - \frac{2n^2}{R^2} V_n'' + \frac{n^4}{R^4} V_n = -\frac{1}{R} \left[-\frac{(2 + v)n}{R} + \frac{n^3 W_n}{R^3} \right] \qquad (10.119)$$

$$W_n^{VIII} - \frac{4n^2}{R^2} W_n^{VI} + \frac{6n^4}{R^4} W_n^{IV} - \frac{4n^6}{R^6} W_n'' + \frac{n^8}{R^8} W_n + 4\varepsilon^4 W_n^{IV} = 0 . \qquad (10.120)$$

Assuming the homogeneous solution of W_n to be of the form

$$W_n = A_{jn} e^{\lambda_{jn} x} \qquad (10.121)$$

its substitution into (10.120) results in the following characteristic equation which will provide eight roots for each value of n.

$$\left[\lambda_{jn}^2 - \frac{n^2}{R^2} \right]^4 + 4\varepsilon^4 \lambda_{jn}^4 = 0 \quad (j = 1, 2, \ldots, 8) . \qquad (10.122)$$

This can be written as

$$\lambda_{jn}^2 - \frac{n^2}{R^2} = \varepsilon\lambda_{jn} \begin{Bmatrix} 1 + i \\ 1 - i \\ -1 + i \\ -1 - i \end{Bmatrix} \tag{10.123}$$

It is seen that λ_{jn} must be complex. Hence, if λ_{jn} is a root, $\overline{\lambda}_{jn}$, the complex conjugate, must also be a root. Also since (10.120) contains only even powers of λ_{jn}, then $-\lambda_{jn}$ and $-\overline{\lambda}_{jn}$ are also roots. Therefore for each n, all eight roots are obtained if two independent roots are found. The following are taken arbitrarily from (10.123) as the two independent equations.

$$\lambda_{1n}^2 - \frac{n^2}{R^2} = \varepsilon\lambda_{1n}(1 + i) \tag{10.124}$$

$$\lambda_{2n}^2 - \frac{n^2}{R^2} = -\varepsilon\lambda_{2n}(1 + i). \tag{10.125}$$

Solutions are easily obtained.

$$\lambda_{1n} = \frac{(1 + i)}{2}\varepsilon + \frac{1}{2}\sqrt{(1 + i)^2\varepsilon^2 + \frac{4n^2}{R^2}} \tag{10.126}$$

$$\lambda_{2n} = -\frac{(1 + i)}{2}\varepsilon + \frac{1}{2}\sqrt{(1 + i)^2\varepsilon^2 + \frac{4n^2}{R^2}}. \tag{10.127}$$

It is seen that λ_{1n} and λ_{2n} are not the negative, the conjugate, or the conjugate of the negative of each other; hence, the eight roots are easily obtained knowing these two independent roots. The homogeneous solution for W_n can be written as:

$$W_{n_H} = \sum_{j=1}^{\infty} A_{jn}e^{\lambda_{jn}x}. \tag{10.128}$$

Similarly,

$$U_{n_H} = \sum_{j=1}^{\infty} B_{jn}e^{\lambda_{jn}x} \tag{10.129}$$

$$V_{n_H} = \sum_{n=1}^{\infty} C_{jn}e^{\lambda_{jn}x} \tag{10.130}$$

where B_{jn} and C_{jn} are functions of A_{jn} only. Keep in mind that only the particular solutions of (10.118) and (10.119) need be used. Substitution of (10.128) and (10.129) into (10.118) provides the necessary relationship.

$$B_{jn} = A_{jn}\frac{-\dfrac{\lambda_{jn}}{R}\left[\dfrac{n^2}{R^2} - v\lambda_{jn}^2\right]}{\left[\lambda_{jn}^2 - \dfrac{n^2}{R^2}\right]^2}. \tag{10.131}$$

A similar expression is easily obtained for C_{jn}.

It must be remembered that a particular solution for w must also be obtained for (10.117), which will also result in additional portions of the particular solutions for u_0 and v_0 in (10.115) and (10.116).

It is seen that even with the simplified set of equations for cylindrical shells shown here (Donnell equations) under asymmetric loads, computations are generally quite laborious. Additionally, they are also sensitive to inaccuracies and care should be taken in the solution of any particular problem. For example, it is desirable that whenever higher powers of λ_{jn} appear in the equations, (10.124) and (10.125) can be used to reduce the roots to the first power.

It should be emphasized that the Donnell assumptions can be made for shells of other geometries if they are 'shallow'. This is a powerful analytical tool.

10.8. Inextensional Shell Theory

A second simplified shell theory that has utility under some conditions is *inextensional* shell theory. Its characteristics are most easily seen when dealing with a simple geometry, and hence, it is treated here with regard to a cylindrical shell, but it is obviously not restricted to shells of this geometry.

Inextensional shell theory finds application when (1) the load applied is non-axially symmetric and over a small portion of the shell, or (2) when the lateral pressure oscillates in a manner such that

$$\int_0^{2\pi} p(x, \theta)\, \mathrm{d}\theta \approx 0 \tag{10.132}$$

and (3) when the ends of the shell are free to deform.

In these cases the shell resists the load primarily in circumferential bending rather than through in-plane (membrane) action. Consider the two following cases, looking at the cross-section of a cylindrical shell.

In the first case, A, the load is uniform around the circumference. The original mid-surface plane (solid line) is displaced uniformly radially to the position denoted

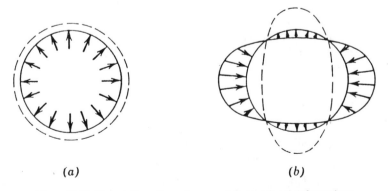

(a) (b)

Figure 10.10. Description of membrane and inextensional deformations.

by the dotted line. No bending occurs at all; there has been an extension of the mid-surface plane (the neutral surface), and this response is called *MEMBRANE* action; it will be discussed in the next section.

Conversely, in the second case the sign of the load oscillates around the circumference, such that the shell mid-surface plane is deformed into an oval shape, in which there is considerable bending, as seen by the change in the curvature, and no extension of the mid-surface circumferential length. Such deformations are therefore denoted as *INEXTENSIONAL*.

Stated in another way a shell deforms under a given load in a manner that results in minimum strain energy. For nonaxially symmetric loads, bending deformations more easily occur than do membrane deformations.

In inextensional shell theory, mid-surface in-plane strains are zero.

$$\varepsilon_x^0 = \varepsilon_{x\theta}^0 = \varepsilon_\theta^0 = 0 . \tag{10.134}$$

These are the only assumptions. The ramifications are now investigated for a circular cylindrical shell. From circular shell theory these mid-surface strains are shown below equated to zero:

$$\varepsilon_x^0 = \frac{\partial u_0}{\partial x} = 0 \tag{10.135}$$

$$\varepsilon_\theta^0 = \frac{1}{R}\left[\frac{\partial v_0}{\partial \theta} + w\right] = 0 \tag{10.136}$$

$$\varepsilon_{x\theta}^0 = \frac{1}{2}\left(\frac{\partial v_0}{\partial x} + \frac{1}{R}\frac{\partial u_0}{\partial \theta}\right) = 0 . \tag{10.137}$$

Substituting these into (10.10) through (10.12), the most important aspects of inextensional shell theory are lucidly seen.

$$N_x = N_\theta = N_{x\theta} = 0 . \tag{10.138}$$

Furthermore from (10.135), $u_0 = u_0(\theta)$ only, and from (10.136), it is seen that

$$w = -\frac{\partial v_0}{\partial \theta} . \tag{10.139}$$

For the solution of a complete circular cylindrical shell, the displacements can again be expanded in a Fourier series in the circumferential direction, as in (10.96). For completeness they are written as follows:

$$u_0 = \sum_{n=1}^{\infty} U_n(x)\cos n\theta + \sum_{m=1}^{\infty} \overline{U}_m \sin m\theta$$

$$v_0 = \sum_{n=1}^{\infty} V_n(x)\sin n\theta + \sum_{m=1}^{\infty} \overline{V}_m(x)\cos m\theta \tag{10.140}$$

$$w_0 = \sum_{n=1}^{\infty} W_n(x)\cos n\theta + \sum_{m=1}^{\infty} \overline{W}_m(x)\sin m\theta .$$

From (10.135), it is seen that u_n and $\bar{u}_m$ are constants. From (10.137), the following relations are found:

$$V_n(x) = \frac{n}{R} U_n x + A_n$$

$$\bar{V}_m(x) = -\frac{m}{R} \bar{U}_m x + B_m$$

(10.141)

where A_n and B_m are constants. Finally, from (10.136), it is seen that

$$W_n(x) = -nV_n(x) = -n\left(\frac{n}{R} U_n x + A_n\right)$$

$$W_m(x) = m\bar{V}_m(x) = m\left(-\frac{m}{R} \bar{U}_m x + B_m\right).$$

(10.142)

Having the expressions for the displacement relations, attention is now turned to the stress couples and shear resultant expressions. Noting from (10.142) that $d^2w/dx^2 = 0$, (i.e., no curvature in the axial direction), (10.21) and (10.22) are both simplified. Using (10.140) through (10.142), the stress couples can be written explicitly as follows:

$$M_\theta = -\frac{D}{R^2} \sum_{n=1}^{\infty} \left[\frac{n^2}{R} U_n x + nA_n\right](n^2 - 1)\cos n\theta$$

$$-\frac{D}{R^2} \sum_{m=1}^{\infty} \left[\frac{m^2}{R} \bar{U}_m x - mB_m\right](m^2 - 1)\sin m\theta$$

(10.143)

$$M_x = vM_\theta$$

(10.144)

$$M_{x\theta} = -\frac{(1 - v)D}{2R^2} \left\{\sum_{n=1}^{\infty} (2n^2 - 1)nU_n \sin n\theta - \sum_{m=1}^{\infty} (2m^2 - 1)m\bar{u}_m \cos m\theta\right\}.$$

(10.145)

It is noted that for $m = n = 1$, both $M_x = M = 0$.

The explicit expression for the shear resultants are found by substituting (10.140) through (10.142) into (10.24) and (10.25), the result is:

$$Q_x = -\frac{D}{R^3} \sum_{n=1}^{\infty} n^2\left(n^2 - \frac{(1 + v)}{2}\right) U_n \cos n\theta$$

$$+\frac{D}{R^3} \sum_{m=1}^{\infty} m^2\left(m^2 + \frac{(1 + v)}{2}\right) \bar{U}_m \sin m\theta$$

(10.146)

$$Q_\theta = \frac{D}{R^3} \sum_{n=1}^{\infty} n^3(n + 1)\left(\frac{n}{R} U_n x + A_n\right) \sin n\theta$$

$$+ \frac{D}{R^3} \sum_{m=1}^{\infty} m^3(m-1)\left(\frac{m}{R} \, \overline{U}_m x - B_m\right)\cos m\theta . \tag{10.147}$$

Looking now at the force equilibrium equations, (10.3) through (10.5), the only meaningful one is the latter, which for the inextensional theory is written as

$$\frac{\partial Q_x}{\partial x} + \frac{1}{R} \frac{\partial Q_\theta}{\partial \theta} = -p(x, \theta) . \tag{10.148}$$

It is obvious that having assumed the displacements to be of the form (10.140), any given lateral loading needs to be expanded as follows:

$$p(x, \theta) = \sum_{n=1}^{\infty} P_n(x) \cos n\theta + \sum_{m=1}^{\infty} \overline{P}_m(x) \sin m\theta \tag{10.149}$$

where $P_n(x)$ and $\overline{P}_m(x)$ would be known functions of the axial coordinate. From the form preceding equations, the following important relations are found.

$$P_n(x) = -(n+1) \frac{n^4}{R^4}\left[\frac{n}{R} U_n x + A_n\right]$$

$$\tag{10.150}$$

$$\overline{P}_m(x) = (m-1) \frac{m^4}{R^4}\left[\frac{m}{R} \overline{U}_m x - B_m\right] .$$

Two very important conclusions are deduced from these equations. First, for inextensional shell theory to be applicable, the lateral load $p(x, \theta)$ can at most be a linear function of x. Second, since U_n, A_n, $\overline{U}_m$, and B_m are uniquely determined by the applied load, no unknown constants remain; therefore, in inextensional shell theory no boundary conditions can be satisfied on the ends of the shell (edges where the meridional coordinate is a constant).

Concerning the first conclusion, if the lateral load is nonlinear in x, but is a slowly varying function in x, inextensional shell theory may still be used to obtain an approximate solution. However, in this case it may be preferable to utilize an energy principle to obtain an approximate solution, wherein the inextensional assumptions are incorporated in the strain energy expression.

Concerning the second conclusion, for any given problem one must live with the values of deflections, stress couples and shear resultants resulting on the ends of the shell. It can also be concluded that for a given shell and loading if the deflections and stresses calculated at the end differ markedly from the actual physical situation, then inextensional shell theory should not be used to obtain a solution, which in turn implies that in that particular case the shell does not deform inextensionally.

10.9. Membrane Shell Theory

A third simplified shell theory is called membrane theory. Although having application to shells of any general shape, it is treated here in detail for the case of a cylindrical shell. The fundamental assumption of this theory is that the shell has no bending resistance, i.e., that all loads are resisted purely extensionally. It is obvious that membrane theory cannot be rationally used when discontinuities in the lateral loading occur, since it was shown previously that such discontinuities always result in a significant bending boundary layer.

One important use of membrane theory, however, is that it is a rational means to obtain a particular solution for many loadings for any shell geometry – or a means to obtain part of the shell particular solution.

To obtain the governing equations for membrane shell theory, it is sufficient to merely insert the following expressions into the governing equations:

$$D = 0 . \tag{10.151}$$

For the case of a circular cylindrical shell, substitition of (10.151) into (10.21) through (10.25) results in the following:

$$M_x = M_\theta = M_{x\theta} = Q_x = Q_\theta = 0 . \tag{10.152}$$

For simplicity, in what follows it shall be assumed that surface shear stresses are zero, hence, $q_x = q_\theta = m_x = m_\theta = 0$, because if this is not the case, one would probably not use membrane theory.

From the third equilibrium equation, (10.5), it is seen that

$$N_\theta = p(x, \theta)R . \tag{10.153}$$

Integration of (10.4) results in

$$N_{x\theta} = N_{\theta x} = -\int \frac{\partial p}{\partial \theta} \, dx + f_1(\theta) \tag{10.154}$$

where $f_1(\theta)$ is to be determined by the boundary conditions. Likewise, (10.3) can be integrated to yield the following:

$$N_x = -\frac{1}{R} \int \frac{\partial N_{x\theta}}{\partial \theta} \, dx + f_2(\theta) \tag{10.155}$$

where $f_2(\theta)$ is another function to be specified by the boundary conditions.

Manipulation of (10.10) through (10.12) results in the following convenient form for the displacements in terms of the stress resultants:

$$\frac{\partial u}{\partial x} = \frac{1}{Eh} [N_x - vN_\theta] \tag{10.156}$$

$$\frac{1}{R} \frac{\partial v_0}{\partial \theta} + \frac{w}{R} = \frac{1}{Eh} [N_\theta - vN_x] \tag{10.157}$$

$$\frac{1}{R} \frac{\partial u_0}{\partial \theta} + \frac{\partial v_0}{\partial x} = \frac{2(1 + v)}{Eh} N_{x\theta} \tag{10.158}$$

Integrating these, expressions for the deflections are found:

$$Ehu_0 = \int (N_x - vN_\theta) \, dx + F_3(\theta) \tag{10.159}$$

$$Ehv = \int 2(1 + v)N_{x\theta}\,dx - \int \frac{Eh}{R}\frac{\partial u_0}{\partial v}\,dx + f_4(\theta) \qquad (10.160)$$

$$Ehw = (N_\theta - vN_x)R - Eh\frac{\partial v_0}{\partial \theta}\,. \qquad (10.161)$$

In the above $f_3(\theta)$ and $f_4(\theta)$ are also functions to be determined by the boundary conditions.

Thus, for a circular cylindrical shell and a given lateral loading $p(x, \theta)$, all stress resultants and displacements are easily found by integrating, in order, (10.153) through (10.155), and (10.159) through (10.161), and subsequently determining the functions f_1 through f_4 by the satisfaction of suitable boundary conditions. It should be noted again that membrane theory solutions for shells of other geometries are equally simple to obtain.

Turning attention to the boundary conditions, it is seen that only four boundary conditions can be satisfied. If the cylindrical shell is closed, then two boundary conditions can be satisfied at each end. From Section 10.1 it is obvious that for $x =$ constant edges the natural boundary conditions are:

Either u_0 prescribed or $N_x = 0$

Either v_0 prescribed or $N_{x\theta} = 0$. $\qquad (10.162)$

Note that in using membrane theory it is possible to specify values of the lateral deflection or the slope at the boundaries. The physical implication of this is that bending stresses and deformations are introduced into the shell whenever the boundary conditions involve restraints on lateral deflections or slopes, such as in the case of simple support or clamped edges, i.e., a 'bending boundary layer' is introduced in that case.

Membrane theory is widely used (and misused) in the analysis of shells. Because it is so easy to obtain solutions it is tempting to utilize it whenever possible. All too often in design, shell thicknesses are determined on the basis of membrane theory, and as a result the shell is underdesigned, as pointed out in Section 10.5.1. When the shell is composed of ductile materials local yielding in the region of high bending stresses 'hides' the inadequate design. However, in shells which utilize less ductile materials and/or when allowable stresses are high, determination of shell thickness by the misuse of membrane theory (ignoring bending stresses) can and does lead to catastrophic results.

As stated earlier, one of the very important uses of membrane theory is to obtain particular solutions for use in the general theory for shells of any shape. Since in so many cases the membrane solution is also a solution to the shell governing differential equations it can effectively be used as the particular solution. This will be illustrated in the following section for the case of a cylindrical shell. Whenever a membrane solution is obtained to be used as a particular solution, either f_1 through f_4 can be taken as zero, or they can be found through the satisfaction of the boundary conditions for the problem that the membrane solution can satisfy. The latter alternative is preferable; the former is acceptable.

10.10. Examples of Membrane Theory

10.10.1. Cylindrical Shell Subjected to a Constant Internal Pressure p_0, with Free Ends

The boundary conditions at each end $(x = 0, L)$ are seen to be $u(0) = N_x(L) = v(0) = v(L) = 0$ which results in $f_1 = f_2 = f_3 = f_4 = 0$. From (10.153) through (10.161), the solution is written as follows:

$$N_\theta = Rp_0 \tag{10.163}$$

$$N_x = N_{x\theta} = 0 \tag{10.164}$$

$$u_0(x) = -\frac{vRp_0x}{Eh} = -\frac{vp_0x}{4\varepsilon^4 RD} \tag{10.165}$$

$$v_0 = 0 \tag{10.166}$$

$$w = \frac{p_0R^2}{Eh} = \frac{p_0}{4\varepsilon^4 D} . \tag{10.167}$$

Note that the solution for u_0 and w given in (10.165) and (10.167) are the particular solutions for the case of $p(x) = p_0$, $N_x = 0$, found previously for the general shell theory given by (10.56) and (10.60). This illustrates the utility of using membrane solutions as particular solutions. Obviously, this example is the simplest imaginable case and membrane theory obviously need not have been employed. However, in more complicated loadings and geometries, the membrane solution used as a particular solution has great advantages when applicable.

10.10.2. Horizontal Cylindrical Shell Filled with a Liquid

Consider the shell shown below, filled with a liquid such that $p(\theta) = p_0 - \gamma R \cos \theta$, where γ is the weight density of the fluid. The shell is clamped at one end and free at the other.

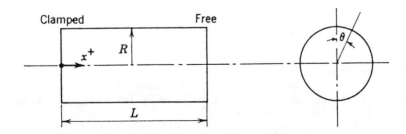

Figure 10.11. Horizontal cylindrical shell filled with a liquid.

Systematic use of equations (10.153) through (10.161) result in the complete solution. The boundary conditions are:

$$u(0) = v(0) = N_x(L) = N_{x\theta}(L) = 0 .$$ (10.168)

The results are as follows:

$$N_\theta = Rp_0 - \gamma R^2 \cos \theta$$ (10.169)

$$N_x = \frac{\gamma}{2} (L - x)^2 \cos \theta$$ (10.170)

$$N_{x\theta} = \gamma R(L - x) \sin \theta$$ (10.171)

$$Ehu_0 = \left[\frac{x^3}{6} + vR^2x - \frac{Lx^2}{2} + \frac{L^2x}{2} \right] \gamma \cos \theta - vRp_0x$$ (10.172)

$$Ehv_0 = 2\gamma R \sin \theta \left[(1 + v) \left(-\frac{x^2}{2} + xL \right) \right.$$

$$\left. + \frac{x^4}{48R^2} + \frac{vx^2}{4} - \frac{Lx^3}{12R^2} + \frac{L^2x^2}{8R^2} \right]$$ (10.173)

$$Ehw = \gamma R \cos \theta \left\{ - R^3 - v(xL - L^2/2) + x^2 - 2xL \right.$$

$$\left. - \frac{x^4}{24R^2} + \frac{x^3L}{6R^2} - \frac{x^2L^2}{4R^2} \right\} + p_0R^2 .$$ (10.174)

Furthermore, the maximum membrane stresses can be found:

$$\sigma_{x_{max}} = \frac{N_x}{n} = \sigma_x \left(0 , \frac{0}{\pi} \right) = \pm \frac{\gamma L^2}{2h}$$ (10.175)

$$\sigma_{\theta_{max}} = \frac{N_\theta}{h} = \sigma_\theta(x, \pi) = \frac{p_0R + \gamma R^2}{h}$$ (10.176)

$$\sigma_{x\theta_{max}} = \frac{N_{x\theta}}{h} = \sigma_{x\theta} \left(0 , \frac{\pi/2}{3\pi/2} \right) = \pm \frac{\gamma RL}{h} .$$ (10.177)

However, it must be remembered that they may be considerably in error from the actual total stresses since in the region $0 \leqslant x \leqslant 4\sqrt{Rh}$, where sizeable bending stresses can be expected to occur.

10.11. References

10.1.   Vinson, J. R., Edge Load Solutions for Conical Shells, *J. of the Engineering Mechanics Division*, ASCE, 92, pp. 37–58, February, 1966.
10.2.   *ASME Boiler and Pressure Vessel Code*, Section III Nuclear Pressure Vessels.
10.3.   Vinson, J. R. and R. L. Sierakowski, *The Behavior of Structures Composed of Composite Materials*, Dordrecht: Martinus-Nijhoff Publishers, 1986.

The following comprise general reference materials on the theory of shells, and are not inclusive, but comprise much of the literature in book form, with most of the engineering solutions that are available today in book form.

10.4. Timoshenko, S. P. and S. Woinowsky-Krieger, *Theory of Plates and Shells*, McGraw-Hill, 1959.

10.5. Novozhilov, V. V., *The Theory of Thin Shells*, Noordhoff, 1959.

10.6. Haas, A. M. and A. L. Bouma, *Proceedings of the Symposium on Shell Research*, North-Holland Publishing Company, 1961.

10.7. Mushtari, Kh. M. and K. Z. Galimov, *Non-Linear Theory of Thin Elastic Shells*, published for NSF and NASA by the Israel Program for Scientific Translations, 1961 (original in Russian dated 1957).

10.8. Ambartsumyan, S. A., *Theory of Anisotropic Shells*, NASA N64-22801, 1964.

10.9. Vlasov, V. Z., *General Theory of Shells and Its Application in Engineering*, NASA TTF-99, N64-19883.

10.10. Gibson, J. E., *Linear Elastic Theory of Thin Shells*, Pergamon Press, 1965.

10.11. Vlasov, V. Z. and N. N. Leont'ev, *Beams, Plates and Shells on Elastic Foundations*, published for NASA and NSF by the Israel Program for Scientific Translations, 1966 (original in Russian dated 1960).

10.12. Ogibalov, P. M., *Dynamics and Strength of Shells*, published for NASA and NSF by the Israel Program for Scientific Translations, 1966 (original in Russian dated 1963).

10.13. Kraus, H., *Thin Elastic Shells*, Wiley, 1967.

10.14. Baker, E. H., A. P. Cappelli, L. Kovalevsky, F. L. Rish, and R. M. Verette, *Shell Analysis Manual*, NASA CR-912, April 1968.

10.15. Baker, E. H., L. Kovalevsky, and F. L. Rish, *Structural Analysis of Shells*, McGraw-Hill, 1972.

10.12. Problems

10.1. A long cylindrical pressure vessel is rated to operate up to 100 psi internal pressure. The wall thickness is $0.1''$ and the mean radius is $50''$. The material is steel with $E = 30 \times 10^6$ psi, $v = 0.3$, and $\sigma_{all} = +120\,000$ psi. One end of the cylindrical vessel is considered clamped. Determine σ_x and σ_θ *at the clamped end on the outer surface.*

10.2. A very long steel pipe ($E = 30 \times 10^6$ psi, $v = 0.3$, $\sigma_{all} = +30\,000$ psi) is to be lifted by using straps located away from either end, tightening up the straps, and subsequently lifting the pipe through hooks attached to the straps. Tightening the straps results in the applied lifting load introducing an axially symmetric radial line load on the pipe which will have some value in pounds per inch of circumference. If the allowable stress is not to be exceeded what is the maximum value of the line load? The shell radius is $12''$ and the wall thickness is $0.2''$.

10.3. In the shell problem of pages 135 and 136, in the bending boundary layer at the simply supported end, $\sigma_{x_{max}}$ occurs at $x = \pi/4\varepsilon$. For the same shell, with $v = 0.3$, calculate $\sigma_\theta(+h/2)$ and $\sigma_\theta(-h/2)$ at the same location, i.e., $x = \pi/4\varepsilon$.

10.4. A lightweight instrument canister to be employed in an orbital research mission consists of a cylindrical shell, with ends which are circular plates, assembled such that at the junction of the cylinder with the circular plate, right angles will continue even under the internal pressures required to prevent electrical arcing. Using the notation of Chapters 6 and 10, what are the boundary conditions at the junction? Use subscripts x for shell and p for the plates.

10.5. The deep submergence instrumentation capsule shown below is designed to provide scientific information at a depth of 1000 feet below the surface of the ocean, where the external pressure is approximately 450 psi. The pressure in the interior of the shell can be considered zero for calculation purposes. The cylindrical shell can be considered as clamped to rigid end plates. The capsule is made of steel ($E = 30 \times 10^6$ psi, $v = 0.3$), is 1 inch thick, 50 inch radius and 100 inches long.

(a) At the end $x = 0$ determine σ_x and σ_θ on both the inside and outside of the shell.

(b) At the end $x = L$ determine σ_x and σ_θ on both the outside and the inside of the shell. (Think!)

(c) At the mid-point of the shell length ($x = L/2$) determine the values of σ_x and σ_θ.

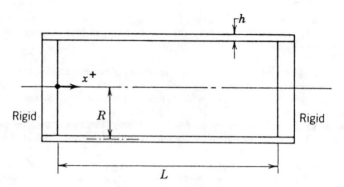

Figure 10.12. Deep submergence capsule.

10.6. The water tower shown at the left is filled to the top such that the pressure exerted on the shell walls is given by $p(x) = \rho(L - x)$ where ρ is the weight density of water. The top of the tank ($x = L$) is considered as a free edge, and the bottom of the tank ($x = 0$) is considered to be clamped. The tank height is considerably longer than $4\sqrt{Rh}$.
(a) What are the boundary value constants M_0, Q_0, M_L, and Q_L?
(b) What is the magnitude and location of the maximum value of σ_x? or σ_θ?
(c) What is the magnitude of the deflection at the top of the tank?

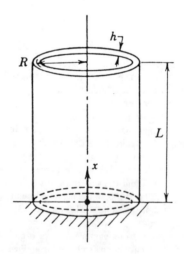

Figure 10.13. Water tower.

10.7. It has been that a bending boundary layer occurs when there is a load or geometric discontinuity (such as an edge). It also occurs if there is a material discontinuity or an abrupt change in wall thickness. Consider a vertical column supporting a Texas tower in which one material is used which will be underwater and extending upward to the height of the highest wave predicted; and a second, cheaper material extending from there to the deck. Thus cylindrical shells of identical geometries but different materials are joined together. If the column is filled with a liquid such as oil or water that there is at the joint an internal pressure p_i, considered constant in the joint area for this calculation, and if the Poisson's ratios of the two materials are equal $v_1 = v_2$, what is the value of the stress couple and the transverse shear resultant at the joint? See Figure 10.14.

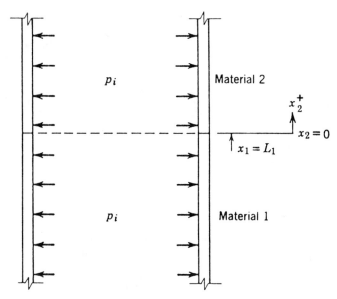

Figure 10.14. Junction of shells with differing materials.

10.8. Consider a long circular cylindrical shell of length L with circular plates as ends such that the shell and the plates are considered clamped at their junction. The shell is subjected to an external pressure, $p_0 = -p_0$ (w constant).

(a) Solve for M_0, Q_0, M_L and Q_L, the boundary value constants for the shell in terms of p_0, ε and v.

(b) If the shell length $L \gg 4\sqrt{Rh}$, what are the stressed σ_x and σ_θ at $x = L/2$.

11

Elastic Stability of Shells

Over and above the determination of stresses and displacements in shell structures, under some loadings the question of elastic stability of a shell arises, just as it does in plates and columns, which were discussed in Chapter 7.

It is beyond the scope of this text to provide an encyclopediatic, in-depth development of buckling of shells. That is the subject of a full length course or courses. The literature treating shell stability is voluminous. Here, only an introduction will be provided. It is intended here, however, to provide sufficient treatment that critical loads or critical stresses can be calculated for cylindrical shells under some common loadings.

One important fact should be remembered. With columns and plates, the critical buckling loads and stresses obtained by analysis prove generally accurate compared to the values obtained experimentally. In the case of shells, the predicted values are never obtained experimentally. Analytical procedures provide a form of solution, but for accuracy, invariably an empirical numerical factor must be used for design and/or accurate analysis.

11.1. Buckling of Isotropic Circular Cylindrical Shells under Axially Symmetric Axial Loads

Consider the governing differential equation for a circular cylindrical shell under axially symmetric loads, derived in Chapter 10, and given in Equations (10.44) and (10.45):

$$\frac{d^4 w}{dx^4} + 4\varepsilon^4 w = \frac{1}{D}\left[p(x) - \frac{\nu N_x}{R} \right] \tag{11.1}$$

where

$$\varepsilon^4 = 3(1 - \nu^2)/h^2 R^2 . \tag{11.2}$$

As in the buckling of columns, the axial critical buckling load does not depend upon the lateral distributed loading, $p(x)$, hence, for calculating the critical load here,

$p(x) = 0$. Therefore, the solution to Equation (11.1) can be written as

$$w(x) = Ae^{-\varepsilon x}\cos\varepsilon x + Be^{-\varepsilon x}\sin\varepsilon x + Ce^{\varepsilon x}\cos\varepsilon x$$

$$+ Ee^{\varepsilon x}\sin\varepsilon x - \frac{vN_xR}{Eh}. \tag{11.3}$$

It is therefore seen that independent of the boundary conditions, i.e., an unrestrained shell, the load N_x causes a lateral displacement, as discussed in the paragraph after Equation (10.45).

$$w = -vN_xR/Eh. \tag{11.4}$$

For the case of buckling, we wish to measure the lateral displacement after the uniform compressive load is applied, not from the unstrained middle surface; hence, defining w as,

$$\overline{w} = w + \frac{vN_xR}{Eh} \tag{11.5}$$

the substitution of Equation (11.5) into (11.1) results in

$$\frac{d^4\overline{w}}{dx^4} + 4\varepsilon^4\overline{w} = 0. \tag{11.6}$$

Through this, one has cancelled out the effects of the axial load (below the critical load) on the lateral displacement, and can now seek under what loading conditions (if any) might other lateral deflections occur due to the axial load N_x.

Recalling the expressions for the governing equations of plates with an in-plane load N_x that causes lateral deflections such as (7.3), and the equation for a beam column given by (7.4), one can analogously modify Equation (11.6) to become

$$\frac{d^4\overline{w}}{dx^4} + 4\varepsilon^4\overline{w} - \frac{N_x}{D}\frac{d^2\overline{w}}{dx^2} = 0. \tag{11.7}$$

Again as was done in determining the buckling loads of beam-columns or plates, one can assume the buckled shape (the mode shape or eigenfunction) for a specified set of boundary conditions and solve for the particular load (the critical load or eigenvalue) that will cause that buckling mode. In this case, if the cylindrical shell is simply supported at both ends, the lateral deflection that will satisfy the boundary conditions is simply supported at both ends, the lateral deflection that will satisfy the boundary conditions is

$$\overline{w}(x) = A\sin\frac{n\pi x}{L}. \tag{11.8}$$

Substituting this expression into (11.7) results in:

$$\left[\frac{n^4\pi^4}{L^4} + 4\varepsilon^4 + \frac{N_x}{D}\frac{n^2\pi^2}{L^2}\right]A\sin\frac{n\pi x}{L} = 0. \tag{11.9}$$

For this to be an equation, it is required that

$$N_x = -D\left[\frac{n^2\pi^2}{L^2} + \frac{EhL^2}{DR^2n^2\pi^2}\right]. \tag{11.10}$$

This means that where any axial load, N_x, will cause a lateral displacement given by Equation (11.4), $w = -\nu N_x R/Eh$, only when the value of axial load is given by Equation (11.10) will the additional lateral displacement given by Equation (11.8) occur. Just as for plates and columns, this latter condition is termed buckling, or elastic instability. For each valve of n, there is a unique buckling mode shape and a unique buckling load. As in all continuous elastic bodies, there are an infinity of mode shapes and buckling loads. However, physically speaking, in a specified elastic body with specified boundary conditions, for a given loading condition in which the load increases from zero, when the lowest buckling load is reached, the body will buckle and most likely will permanently deform or fracture. Thus, in buckling one usually seeks to find the lowest buckling load and mode shape for a given loading mechanism, and there is little interest in higher loads and modes. This is one difference between buckling and natural vibration problems, where in the latter one is usually interested in determining a large number of vibration modes (eigenfunctions) and natural frequencies (eigenvalues).

Pursuing this example problem further, letting $n = 1$ in Equation (11.10) results in

$$N_{x_{cr}} = -D\left[\frac{\pi^2}{L^2} + \frac{EhL^2}{DR^2\pi^2}\right]. \tag{11.11}$$

It is seen that N_x is a function of the length of the shell, L, and to find the minimum buckling load, N_x, we set the derivative of N_x with respect to L equal to zero. The length at which the minimum buckling load occurs is

$$L = \pi\left[\frac{R^2h^2}{12(1-\nu^2)}\right]^{1/4} \tag{11.12}$$

and the minimum buckling load is

$$N_{x_{min}} = -\frac{Eh^2}{R\sqrt{3(1-\nu^2)}}. \tag{11.13}$$

Note that if L is less or greater than the value given by (11.12), the buckling load will be higher for $n = 1$.

Returning to Equation (11.10) and letting $n = 2, 3, 4, \ldots$ and following the same procedure, in each case the minimum buckling load equals that given by Equation (11.13). Hence, it is conservative to say that for axially symmetric buckling of a simply supported, circular cylindrical, isotropic shell under axial loading

$$N_{x_{cr}} = -\frac{Eh^2}{R\sqrt{3(1-\nu^2)}} \quad \text{for} \quad L \geqslant \pi\left[\frac{R^2h^2}{12(1-\nu^2)}\right]^{1/4} \tag{11.14}$$

or

$$\sigma_{x_{cr}} = \frac{N_{x_{cr}}}{h} = -\frac{Eh}{R\sqrt{3(1-v^2)}} \; . \tag{11.15}$$

In some literature where $v = 0.3$, one often sees the relation written as

$$\sigma_{x_{cr}} = -0.605 \frac{Eh}{R} \quad \text{for} \quad L \geqslant 1.72\sqrt{Rh}. \tag{11.16}$$

As stated previously, in practice these buckling loads cannot be reached. Equations (11.14) and (11.16) represent the critical stress causing buckling using a linear theory. Moreover, where there exists a close correlation between theory and experimental data of elastic and plastic buckling of flat plates and columns under various types of loadings and boundary conditions, no such correlation exists in shells under axial compressive loads. This implies that in the cylindrical shells, initial imperfections are very important, and this has led to a great deal of study in the past. The buckling load is particularly important, because under axial compression buckling is synonymous with collapse of the shell.

Hence, it is necessary to incorporate an empirical factor in all equations in order to relate the theoretical values to the actual test data. From Reference 11.1, Equation (11.15) is modified to become

$$\sigma_{x_{cr}} = -\frac{\gamma E}{\sqrt{3(1-v^2)}} \frac{h}{R} \tag{11.17}$$

where in this case,

$$\gamma = 1 - 0.901(1 - e^{-\phi}) \tag{11.18}$$

and where

$$\phi = \frac{1}{16} \sqrt{\frac{R}{h}} \tag{11.19}$$

The coefficient γ, based on Reference 11.2, insures a good lower bound on the major portion of existing test data. However, References (11.1) and (11.2) state that these expressions should not be used for $L/R \geqslant 5$, because of insufficient experimental verification in that range. However, for preliminary design they can be used in lieu of any other expressions available.

Moreover, because of the difficulties in characterizing analytically end restraints greater than those of simple support, Equation (11.17) should be used unless there is specific experimental verification on the particular shell geometry and boundary conditions, which would allow a larger value of γ to be used.

In using (11.17) through (11.19) to analyze what the critical stress or critical load is for a shell of specified geometry and material, the procedure is straightforward. However, note that in designing a shell, i.e., determining the thickness h, for a shell to carry a particular load without buckling, the procedure becomes an iterative one.

Note also that the total load P_{cr} (lbs) corresponding to the critical stress, $\sigma_{x_{cr}}$, of (11.17) is

$$P_{cr} = 2\pi Rh\sigma_{x_{cr}}.\tag{11.20}$$

Geometrically speaking, the buckle pattern that occurs is usually that of a diamond shape of dimensions small compared to the circumference of shell length.

11.2. Buckling of Isotropic Circular Cylindrical Shells under Axially Symmetric Axial Loads and an Internal Pressure

Lo, Crate, and Schwartz, Reference 11.3, studied this problem in some detail, and showed that buckling loads increase due to the effects of internal pressure. They postulate that the increase is due to the membrane effects of the internal pressure reducing the effects of local imperfections in the shell.

Their results can be plotted as follows, where C is the coefficient in the equation below, and p is the internal pressure.

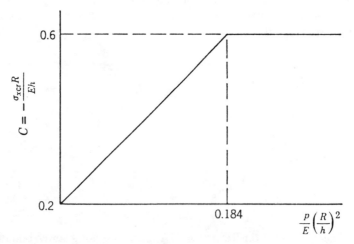

Figure 11.1. Effect of internal pressure on the buckling of a cylindrical shell subjected to axial compressive loads according to Reference 11.3.

Thus, when $p \geqslant 0.184\, E(h/R)^2$, the classical buckling stress of Equation (11.15) or (11.16) is reached. To generalize the results of Reference 11.3, calculate the ordinate value from Equations (11.17) through (11.19) to obtain the $p = 0$ value (not necessarily 0.2). Then interpolate linearly between that value and the classical value for an abscissa at 0.184.

11.3. Buckling of Isotropic Circular Cylindrical Shells under Bending

Again, buckling and collapse coincide for isotropic unpressurized circular cylinders in bending. Equations (11.17) through (11.19) can again be used but based upon the data in Reference 11.2, because there is not good agreement between experiment and theory:

$$\gamma = 1 - 0.731(1 - e^{-\phi}) \qquad (11.21)$$

where

$$\phi = \frac{1}{16} \sqrt{\frac{R}{h}} . \qquad (11.22)$$

In this case the buckling stress of Equation (11.17) is related to the overall bending moment by

$$\sigma = \frac{MR}{\pi R^3 h} = \frac{M}{\pi R^2 h}$$

where $I = \pi R^3 h$.

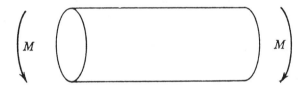

Figure 11.2. Circular cylindrical shell subjected to a bending moment.

As an axial compression, the buckle pattern is diamond shape, initially occurring along the axial generator associated with the highest compressive stress.

11.4. Buckling of Isotropic Circular Cylindrical Shells under Lateral Pressures

It is found that a long cylindrical shell under lateral or hydrostatic loading will buckle into two circumferential harmonic waves in the same manner as a ring. As the cylinder length decreases, the number of circumferential waves will increase with a consequent increase in buckling stress.

For very short cylinders under lateral pressure, the behavior corresponds to that of a long flat plate under longitudinal compression, with the boundary conditions along its longitudinal edges corresponding to those along the cylinder edges.

The hydrostatically loaded short cylinder theoretically approaches that of a long flat plate with biaxial compressive loading where in the sketch above the long edges would have an applied stress of $\sigma = pR/2h$, in addition to the stress shown on the short edge.

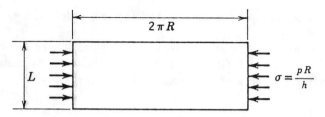

Figure 11.3. Analysis of buckling due to external pressure.

Thus in each of the above cases, the methods of Chapters 7 can be employed.

For a long shell, defined as $\gamma Z > 100$, where $z = L^2(1 - v^2)^{1/2}/Rh$, the critical pressure causing buckling is

$$p_{cr} = \frac{0.926\sqrt{\gamma}E}{(R/h)^{5/2}(L/R)} \ . \tag{11.23}$$

Available experimental data agree reasonably well with classical buckling theory, but some tests fall as much as five percent below theory, as shown in Reference 11.4. Hence, Reference 11.1 recommends using $\gamma = 0.90$ with Equation (11.23).

11.5. Buckling of Isotropic Circular Cylindrical Shells in Torsion

Here, the agreement between linear theory and experimental results for the buckling of cylindrical shells in torsion is much better than the same cylinders subjected to axial compression or bending. This means that under torsional loadings initial imperfections are relatively unimportant.

The analytical methods are more complicated than those for axial loads or lateral pressures because the assumed mode shape functions are not products of trigonometric functions. Physically, this means that there are no generators which remain straight during buckling. In torsion the nodal lines are of a helical shape.

Other differences noted are that in buckling under torsion in the elastic range, buckling is not accompanied by immediate collapse. Collapse of an isotropic circular cylinder in torsion occurs at a considerable higher twist angle. However, the collapse load is only slightly higher than the buckling load, which should be used as a good approximation for the collapse load in design. For geometries of $50 < \gamma z < 78(R/h)^2 (1 - v^2)$ where $z = L^2(1 - v^2)/Rh$

$$\sigma_{x\theta cr} = \frac{0.747\,\gamma^{3/4}E}{(R/h)^{5/4}(L/R)^{1/2}} \ . \tag{11.24}$$

For $\gamma z > 78(R/n)^2 (1 - v^2)$

$$\sigma_{x\theta cr} = \frac{\gamma E}{3\sqrt{2}(1 - v^2)^{3/4}} \left(\frac{h}{R}\right)^{3/2} \ . \tag{11.25}$$

In each, $\gamma = 0.80$ is recommended in Reference 11.1 to approximate the lower limit of the bulk of the experimental data. In each case, the $\sigma_{x\theta_{cr}}$ is related to an overall torque by the following:

$$T = 2\pi R^2 h \sigma_{x\theta}.$$

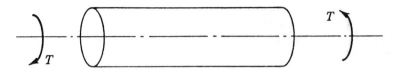

Figure 11.4. Circular cylindrical shell subjected to a torsional load *T*.

11.6. Buckling of Isotropic Circular Cylindrical Shells under Combined Axial Loads and Bending Loads

Since the nature of the buckle patterns is the same for axial compression and bending, a linear interaction curve fits all test data well:

$$R_c + R_B = 1$$

where

$$R_c = \frac{\sigma_x}{\sigma_{x_{cr}}}, \quad R_B = \frac{\sigma_B}{\sigma_{B_{cr}}}$$

$$\sigma_x = N_x/h = P/2\pi Rh \tag{11.26}$$

$$\sigma_B = \frac{MR}{\pi R^3 h}$$

[Where the negative value of the stress value is used.]

The stress couple *M* is the total stress couple applied to the shell in a beam bending type context, and is given in terms of in.–lbs. $\sigma_{x_{cr}}$ and $\sigma_{B_{cr}}$ above are given by Equations (11.17), using (11.18) and (11.21) respectively.

Using Equation (11.26), the buckling occurs when a combination of applied axial load and bending load causes the ratios to be large enough to equal unity.

11.7. Buckling of Isotropic Circular Cylindrical Shells under Combined Axial Load and Torsion

In this case, use

$$R_c + R_T^2 = 1 \tag{11.27}$$

where

$$R_T = \frac{\sigma_{x\theta}}{\sigma_{x\theta_{cr}}} , \quad \sigma_{x\theta} = \frac{T}{2\pi R^2 h} .$$ (11.28)

Here T is a total applied torque to the end of the cylindrical shell, and $\sigma_{x_{cr}}$ is given by Equation (11.24) or (11.25).

11.8. Buckling of Isotropic Circular Cylindrical Shells under Combined Bending and Torsion

Here the interaction equation can be written

$$R_B^{1.5} + R_T^2 = 1$$ (11.29)

where all terms have been defined previously.

11.9. Buckling of Isotropic Circular Cylindrical Shells under Combined Bending and Transverse Shear

In the case where a shell is subjected to the action of a stress couple M and a transverse shear resultant V, in the context of a beam, as shown below the interaction equation is found to be

$$R_B + R_s^2 = 1$$

where

$$R_s = \frac{\sigma_{x\theta_{max}}}{\sigma_{T_{cr}}*}$$ (11.30)

and

$$\sigma_{x\theta_{max}} = V/\pi R h$$

$$\sigma_{T_{cr}}* = 1.25\sigma_{x\theta_{cr}} .$$

In the above $\sigma_{x\theta_{cr}}$ is given by either Equation (11.24) or (11.25).

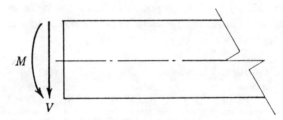

Figure 11.5. Circular cylindrical shell subjected to a bending moment M and a transverse shear resultant.

11.10. Buckling of Isotropic Circular Cylindrical Shells under Combined Axial Compression, Bending and Torsion

Here the interaction equation can be written as:

$$R_c + R_B + R_T^2 = 1 .$$ (11.31)

This is a tentative relationship until more test data becomes available.

11.11. Buckling of Isotropic Spherical Shells under External Pressure

Here also there s much evidence toward premature buckling of spherical shells due to imperfect sphericity, and this effect is even more pronounced in spherical shells than is cylindrical shells.

An empirical buckling equation developed at the Naval Ship Research and Development Center provides a conservative estimate of the critical buckling external pressure as,

$$p_{cr} = -0.84E \left(\frac{h}{R}\right)^2$$ (11.32)

where, of course, $\sigma = pR/2h$. This equation is satisfactory whenever initial departures from sphericity are less than $2\frac{1}{2}\%$ of the shell thickness.

11.12. Buckling of Anisotropic and Sandwich Cylindrical Shells

Pertinent equations for orthotopic cylindrical shells and isotropic sandwich cylindrical shells are found in Reference 11.1.

The buckling of stiffened cylinders is complicated by the fact that both overall instability and local instability can offer. This is dealt with in Reference 11.5.

11.13. References

11.1. Anon, *Buckling of Thin-Walled Circular Cylinders*, NASA SP-8007, September, 1965.
11.2. Seide, P., V. I. Weingarten, and E. J. Morgan, *The Development of Design Criteria for Elastic Stability of Thin Shell Structures*, STL/TR-60-0000-19425 (AFBMO/TR-61-7). Space Technology Laboratory, Inc., December 31, 1960.
11.3. Lo, H., H. Crate, and E. B. Schwartz, *Buckling of Thin Walled Cylinders Under Axial Compression and Internal Pressure*, NACA TN 1027, 1951.
11.4. Batdorf, S. B., *A Simplified Method of Elastic Stability Analysis for Thin Cylindrical Shells*, NACA Report 874, 1947.
11.5. Becker, H., *Handbook of Structural Stability, Part VI, Strength of Stiffened Curved Plates and Shells*, NACA TN 3786, 1958.

11.14 Problems

11.1. Consider a cylindrical interstage structure on a missile system of length $L \geqslant 1.72\,Rh$, yet $L/R \leqslant 5$, composed of magnesium ($E = 6.5 \times 10^6$ psi, $v = 0.3$, $\sigma_{\text{yield}} = +40\,000$ psi. If a shell of 30″ radius and 0.1″ thickness is subjected to an axial compressive load, what will be the critical stress?

11.2. A spherical deep submersible is composed of steel ($E = 30 \times 10^6$ psi, $\sigma_{\text{ult}} = 140\,000$ psi, $v = 0.3$), and has a geometry of 8′ is diameter and 0.5″ wall thickness. If the ocean pressure is given by $p = 0.45\,d$, where p is in psi and d is in feet, to what depth may the submersible go before buckling occurs? What is the wall stress at the maximum depth determined above?

11.3. A long cylindrical section of a missile of 40 inch radius and 1 inch thickness is subjected to both bending and axial compression during launch. If the maximum bending moment is 30 000 000 in.-lbs, what axial load can be tolerated without buckling the aluminum shell? ($E = 10 \times 10^6$ psi, $v = 0.3$, $\sigma_{\text{all}} = \pm 30\,000$ psi.)

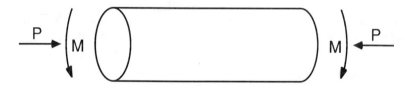

Figure 11.6. Cylindrical shell under bending and axial compression.

11.4. A long aluminum cylindrical shell support for a water tower ($E = 10 \times 10^6$ psi, $v = 0.3$, $\sigma_{\text{all}} = 30\,000$ psi) of 40 inch radius must withstand an axial load of 100 000 lbs without buckling. To the nearest one hundredth of an inch, what shell thickness is required?

11.5. A cylindrical shell is composed of high strength steel having the following properties: $E = 30 \times 10^6$ psi, $\sigma_{\text{allowable}} = 200\,000$ psi, $v = 0.3$, $\rho = 0.283$ lbs./in.^{3} The shell is 3/8″ thick, 25″ radius, and is 50″ long. Calculate the total axial compressive load (in pounds) that the shell can resist before buckling, if the shell is unpressurized.

11.6. What load (in pounds) can the shell resist in Problem 11.5, if the shell were internally pressurized to 700 psi?

11.7. In the shell of Problem 11.5, what is the length of the bending boundary layer?

11.8. For the shell of Problem 11.5, what is the maximum bending moment, M, that the shell can resist without buckling?

11.9. For the shell of Problem 11.5, what is the maximum external pressure that the shell can withstand without buckling?

11.10. For the shell of Problem 11.5, what is the maximum torque T that can be resisted by the shell without it buckling?

11.11. A cylindrical shell is composed of steel which has the following properties: $E = 30 \times 10^6$ psi, $\sigma_{\text{allowable}} = 300\,000$ psi, $v = 0.3$, $\rho = 0.283$ lbs./in.^{3} The shell is 1/4″ thick, 25″ radius, and is 125″ long. Calculate the total axial compressive load (in pounds) that the shell can resist before buckling, when the shell is unpressurized.

11.12. What load can the shell resist, of Problem 11.11, if it was internally pressurized to 600 psi.

11.13. What load would buckle the unpressurized shell of Problem 11.11 as a simply supported column?

11.14. In the shell of Problem 11.11, what is the length of the bending boundary layer?

11.15. For the shell of Problem 11.11, what is the maximum bending moment, M, that the shell can withstand before buckling?

11.16. For the shell of Problem 11.11, what is the maximum external pressure the shell can withstand without buckling?

11.17. For the shell of Problem 11.11, what is the largest torque T that can be applied to the shell without it buckling?

11.18. A cylindrical steel shell with material properties of $E = 30 \times 10^6$ psi, $\sigma_{all} = 200000$ psi, $v = 0.3$, $p = 0.283$ lb./in.^{3}, is $0.1''$ thick, of radius $25''$, and is $150''$ long.

    (a) Calculate the total axial compressive load (lbs.) that the shell can resist without buckling, when the shell is unpressurized.

    (b) What load can the shell above resist without buckling if it is internally pressurized to 500 psi?

    (c) For the shell above, what is the length of the bending boundary layer?

11.19. For the shell of Problem 11.18, what is the maximum M (in.-lbs.) that the unpressurized shell can withstand without buckling?

11.20. A long cylindrical shell is subjected to an axially compressive load of 1×10^6 lbs. and a bending moment $M = 1 \times 60^6$ in.-lbs. If this steel cylinder [30×10^6 psi, $v = 0.3$], of $30''$ radius is designed to resist buckling, what wall thickness is required? Note: determine it only to the nearest $0.1''$.

11.21. A cylindrical shell is composed of steel with the following properties: $E = 30 \times 10^6$ psi, $\sigma_{all} = 150000$ psi, $v = 0.3$, $\rho = 0.283$ lb./in.^{3}. The shell is $1/8''$ thick. Calculate the total axial compressive load (lbs.) that the shell can resist without buckling when the shell is unpressurized.

11.22. What load (lbs.) can the shell of Problem 11.21 resist without buckling if it is internally pressurized to 600 psi?

11.23. For the shell of Problem 11.21, what is the length of the bending boundary layer?

11.24. For the shell of Problem 11.21, what is the maximum bending moment M (in.-lb.) that the unpressurized shell can withstand without buckling?

11.25. For the shell of Problem 11.21, what is the maximum uniform external pressure the shell can withstand without buckling?

11.26. For the shell of Problem 11.21, what is the largest torque, T (in.-lbs.) that can be applied to the shell without it buckling?

11.27. Consider a cylindrical shell with the dimensions $L = 50''$, $R = 20''$, and $h = 0.1''$. The steel shell has properties $E = 30 \times 10^6$ psi, $v = 0.3$, $\sigma_{allowable} = 60000$ psi, $\rho = 0.283$ lb./in.^{3}. It is clamped at each end.

    (a) What is the length of the bending boundary layer at each end?

    (b) If the internal pressure inside the shell is 100 psi, what is σ_x and σ_θ at the shell midlength (i.e., $x = 25''$)?

    (c) If the shell is unpressurized, what is the maximum compressive axial load P (lbs.) that the shell can withstand without buckling?

    (d) At that buckling load, would the shell be overstressed? (*Note*: neglect bending stresses in the bending boundary layer.)

    (e) What internal pressure would be required to obtain the classical buckling stress?

The Vibration of Cylindrical Shells

The vibration of cylindrical shells, as well as shells of more complex shapes, is rather involved, far beyond the scope and intended purpose of this text. It is complicated due to several factors:

a. All but some very simple modes involve non-axially symmetric mode shapes.
b. There are numerous shell theories, each of which differ one from the other.
c. There are several simplifications, such as the Donnell approximations, but it is not clear when they can be made, and obtain accurate results.
d. The curvature of the shell forces a coupling of the differential equations involving the displacements u_0, v_0, and w.
e. Many forcing functions of practical importance are also non-axially symmetric.

Therefore, only the simplest of treatments is included herein to allow the reader to see approaches that can be used in a detailed study. Knowing also that results are important to practicing engineers, the recommendation to utilize the inclusive work of Leissa (Reference 12.1) is hereby made.

12.1. Governing Differential Equations for Natural Vibrations

Looking at Equations (6.12) through (6.23), and using d'Alembert's Principle, it is seen that the governing equations to be used in studying the dynamic response for cylindrical shells are:

$$\frac{\partial^2 u_0}{\partial x^2} + \frac{(1-v)}{2}\frac{\partial^2 u_0}{\partial s^2} + \frac{(1+v)}{2}\frac{\partial^2 v_0}{\partial x \partial s} + \frac{v}{R}\frac{\partial w}{\partial x} = +\frac{\rho h}{K}\frac{\partial^2 u_0}{\partial t^2} \tag{12.1}$$

$$\left[\frac{(1-v)}{2}\frac{\partial^2 v_0}{\partial x^2} + \frac{\partial^2 v_0}{\partial s^2}\right] + \frac{1}{R}\frac{\partial w}{\partial s} + \frac{(1+v)}{2}\frac{\partial^2 u_0}{\partial x \partial s} - k^2 R\frac{\partial}{\partial s}(\nabla^2 w) = +\frac{\rho h}{K}\frac{\partial^2 v_0}{\partial t^2} \tag{12.2}$$

$$\nabla^4 w - \frac{1}{R}\frac{\partial}{\partial s}(\nabla^2 v_0) + \frac{1}{k^2}\left[\frac{1}{R^3}\frac{\partial v_0}{\partial s} + \frac{w}{R^4} + \frac{v}{R^3}\frac{\partial u_0}{\partial x}\right] = \frac{p(x,s,t)}{D} - \frac{\rho h}{D}\frac{\partial^2 w}{\partial t^2}. \tag{12.3}$$

Of course, separation of variables can be attempted for the case of obtaining natural frequencies (i.e., $p = 0$). The circumferential portion of the solution is repeatable for every 2π radians, and the time function is harmonic. Therefore a solution for u_0, v_0, and w would include:

$$\begin{Bmatrix} u_0 \\ v_0 \\ w \end{Bmatrix} = \sum_{m=1}^{\infty} \sum_{n=1}^{\infty} \begin{Bmatrix} U(x) \cos \omega_{mn} t \\ V(x) \sin \omega_{mn} t \\ W(x) \cos \omega_{mn} t \end{Bmatrix}. \tag{12.4}$$

This would result in a set of ordinary differential coupled equations in the variables U, V and W, for which solutions can be found for various boundary conditions. Formally this can be done, and Leissa provides many solutions.

12.2. Hamilton's Principle for Determining the Natural Vibrations of Cylindrical Shells

As an alternative, the Theorem of Minimum Potential Energy and Hamilton's Principle can be used to determine natural vibration frequencies for cylindrical shells.

In using Minimum Potential Energy for this purpose, the potential energy for the shell is the total strain energy since no surface tractions are imposed. The strain energy, U, for the cylindrical shell is found straightforwardly from Equation (9.1) and (9.4), and the results are given below without derivation.

$$\begin{aligned}
U = \int_0^{2\pi R} \int_0^L \Bigg\{ &\frac{Eh}{2(1-v^2)} \Bigg[\left(\frac{\partial u_0}{\partial x} \right)^2 + \frac{\partial v_0}{\partial s} + 2 \frac{w}{R} \frac{\partial v_0}{\partial s} + \left(\frac{w}{R} \right)^2 \Bigg] \\
&+ \frac{Eh^3}{24(1-v^2)} \Bigg[\frac{\partial^2 w}{\partial x^2} + \frac{1}{R^2} \left(\frac{\partial v_0}{\partial s} \right)^2 - \frac{2}{R} \frac{\partial v_0}{\partial s} \frac{\partial^2 w}{\partial s^2} + \left(\frac{\partial^2 w}{\partial s^2} \right)^2 \Bigg] \\
&+ \frac{vEh}{(1-v^2)} \Bigg[\frac{\partial u_0}{\partial x} \frac{\partial v_0}{\partial s} + \frac{w}{R} \frac{\partial u_0}{\partial x} \Bigg] \\
&+ \frac{vEh^3}{12(1-v^2)} \Bigg[-\frac{1}{R} \frac{\partial^2 w}{\partial x^2} \frac{\partial v_0}{\partial s} + \frac{\partial^2 w}{\partial x^2} \frac{\partial^2 w}{\partial s^2} \Bigg] \\
&+ \frac{Eh}{(1-v)} \Bigg[\frac{1}{4} \left(\frac{\partial u_0}{\partial s} \right)^2 + \frac{1}{2} \frac{\partial u_0}{\partial s} \frac{\partial v_0}{\partial x} + \frac{1}{4} \left(\frac{\partial v_0}{\partial x} \right)^2 \Bigg] \\
&+ \frac{Eh^3}{12(1+v)} \Bigg[\frac{1}{4R^2} \left(\frac{\partial v_0}{\partial x} \right)^2 - \frac{1}{R} \frac{\partial^2 w}{\partial x \partial s} \frac{\partial v_0}{\partial x} + \left(\frac{\partial^2 w}{\partial x \partial s} \right)^2 \Bigg] \Bigg\} \, dx \, ds. \tag{12.5}
\end{aligned}$$

If one is interested primarily in the radial vibrations, which often is the case, the kinetic energy T can be approximated as

$$T = \int_A \frac{1}{2} \rho h \left(\frac{\partial w}{\partial t}\right)^2 \tag{12.6}$$

where all symbols have been defined previously.

Then, approximate solutions can be obtained by letting the displacement take the following form for shell simply supported at each end, for example:

$$w(x, s, t) = \sum_{m=1}^{\infty} \sum_{n=1}^{\infty} A_{mn} \sin \frac{m\pi x}{L} \sin n\theta \sin \omega_{mn} t . \tag{12.7}$$

Suitable expressions for the in-place displacement U_0 and V_0 need to be formulated, depending upon the problem studied. The result will be a spectrum of frequencies ω_{mn}, from which the lowest one will be the fundamental frequency. Also, the analytical results will become increasingly erroneous for higher values of m and n, because when the wavelength of vibration approaches the thickness of the shell, transverse shear deformation should be included which will modify the problem. The effects of transverse shear deformation are the lowering of the natural frequencies from the values predicted by the classical solution.

12.3. Reference

12.1. Leissa, A. W., *Vibration of Shells*, NASA SP 288, 1973.

APPENDIX 1
Properties of Useful Engineering Materials

The following Table provides representative material properties necessary to perform analysis and design calculations of structures using methods provided throughout the text. The properties are those for ambient temperature (70 °F).

The symbols in the Table are as follows:

ρ density (lbs./in.^{3})
E_T tensile modulus of elasticity (lbs./in.^{2})
σ_{uT} ultimate tensile strength (lbs./in.^{2})
σ_{yr} tensile yield strength (lbs./in.^{2})
σ_{yc} compressive yield strength (lbs./in.^{2})
σ_s shear strength (lbs./in.^{2})
ε_F elongation to failure (%)
α coefficient of thermal expansion (in./in./°F).

All of the metallic material properties are taken from *Materials Engineering*, Materials Selector Issue, a Reinhold Publication.

Representative Material Properties

Material	ρ	E_r	σ_{uT}	σ_{yT}	ε_F	α	σ_{yc}	σ_s
Steels								
A-286	0.286	29.1	146	100	25	10.3		
Stainless W	0.280	28	195	180	3–15	5.5	180	120
17-7 Stainless	0.276	29	200	185	9	5.6	200	136
PH 15-7 Mo	0.277	29	210	200	7	5	217	143
AM 350 Stainless	0.282	29.4	206	173	13.5	6.3		
AISI 301 Austenitic								
Stainless	0.29	28.0	185	140	8–9	9.4		
D-6A	0.283	30	284	250	7.5			
4340	0.283	30	287	270	11	6.3		

171

Material	ρ	E_r	σ_{uT}	σ_{yT}	ε_F	α	σ_{yc}	σ_s
Aluminum Alloys								
2024-T3	0.100	10.6	70	50	18	12.9		41
6061-T6	0.098	10.0	45	40	12	13.0		30
7075-T6	0.101	10.4	83	73	11	13.1		48
7079-T6	0.099	10.3	78	68	14	13.1		45
7178-T6	0.102	10.4	88	78	10–11	13.0		52
Cobalt Superalloy								
Haynes Alloy 25	0.33	34.2	146	67	64	9.4		
Magnesium Alloy								
AZ31B-F	0.064	6.5	36–38	24–28	12–16	14	12–14	19
HK31A-H24	0.065	6.4	37	29		16	23	27
Nickel Alloys								
Monel 400 (Spring)	0.310	26	100–140	25–45	50–35	7.7		
Inconel 610	0.300	31	70–95	30–45	30–10	8.92		
Hastelloy B	0.334	26.4	121	56.5	63			
Udimet 500	0.290	31.2	197	110	18	9.8		
René 41	0.298	31.8	206	154	14	9.3		
Titanium Alloys								
Ti-6A1-4V	0.160	16.5	138	128	12	5.3		100
Ti-7A1-4Mo	0.162	16.2	160	150	16	5.6	140	105

APPENDIX 2:
Answers to Selected Problems

2.1.

$$e = \frac{h}{2}\left[\frac{\partial}{\partial x}\left(\tau_{1x} + \tau_{2x}\right) + \frac{\partial}{\partial y}\left(\tau_{1y} + \tau_{2y}\right)\right]$$

$$f = \frac{1}{1-v}\left[2\frac{\partial^2}{\partial y^2}\left(\tau_{1x} - \tau_{2x}\right) + (1-v)\frac{\partial^2}{\partial x^2}\left(\tau_{1x} - \tau_{2x}\right)\right]$$

$$+ \frac{(1+v)}{(1-v)}\frac{\partial^2}{\partial x\,\partial y}\left(\tau_{1y} - \tau_{2y}\right)$$

$$g = \frac{1}{(1-v)}\left[2\frac{\partial^2}{\partial x^2}\left(\tau_{1y} - \tau_{2y}\right) + (1-v)\frac{\partial^2}{\partial y^2}\left(\tau_{1y} - \tau_{2y}\right)\right]$$

$$+ \frac{(1+v)}{(1-v)}\frac{\partial^2}{\partial x\,\partial y}\left(\tau_{1x} - \tau_{2x}\right)$$

4.1.

$$C_1 = -\frac{A_n}{D\lambda_n^4}, \quad C_3 = -\frac{C_2}{\lambda_n}$$

$$C_2 = -\frac{A_n}{D\lambda_n^3}\frac{[\cosh\lambda_n a - 1 - \lambda_n a]\sinh\lambda_n a + \lambda_n a\cosh\lambda_n a\,(\cosh\lambda_n a - 1)}{\sinh^2\lambda_n a - (\lambda_n a)^2}$$

$$C_4 = \frac{A_n}{D\lambda_n^3}\frac{(1 - \lambda_n a)\sinh\lambda_n a + \lambda_n a\cosh\lambda_n a\,(\sinh\lambda_n a - 1)}{\sinh^2\lambda_n a - (\lambda_n a)^2}$$

4.3.

$$C_1 = -\frac{A_n}{D\lambda_n^4}, \quad C_4 = \frac{A_n}{2D\lambda_n^3}$$

$$C_2 = \frac{A_n}{2D\lambda_n^3}\frac{\sinh\lambda_n a + 2\coth\lambda_n a(1 - \cosh\lambda_n a)}{\cosh\lambda_n a + \lambda_n a\sinh\lambda_n a - \lambda_n a\cosh\lambda_n a\coth\lambda_n a}$$

$$C_3 = \frac{A_n}{2D\lambda_n^4\sinh\lambda_n a}\frac{(\lambda_n a)^2 + 2\cosh\lambda_n a\,(\cosh\lambda_n a - 1) - 2\lambda_n a\sinh\lambda_n a}{\cosh\lambda_n a + \lambda_n a\sinh\lambda_n a - \lambda_n a\cosh\lambda_n a\coth\lambda_n a}$$

4.5. $C_1 = \dfrac{A_n v}{2D \lambda_n^4} \times$

$$\times \left\{ \dfrac{1 + v}{1 - v} \sinh^3 \lambda_n a - \dfrac{2}{v} \left[\lambda_n a \sinh \lambda_n a + \dfrac{2}{(1 - v)} \cosh \lambda_n a \right] \right.$$

$$\left. + (\lambda_n a)^2 \middle/ \dfrac{(3 + v)}{2} \sinh^2 \lambda_n a + \dfrac{1 - v}{2} (\lambda n a)^2 + \dfrac{2}{(1 - v)} \right\}$$

$$C_2 = \dfrac{A_n v}{2D \lambda_n^3} \times \left\{ (1 - \cosh \lambda_n a)(\sinh \lambda_n a - \lambda_n a) \right.$$

$$+ \dfrac{1}{v} (\sinh \lambda_n a - \lambda_n a \cosh \lambda_n a) \middle/ \dfrac{(3 + v)}{2} \sinh^2 \lambda_n a$$

$$\left. + \dfrac{1 - v}{2} (\lambda_n a)^2 + \dfrac{2}{1 - v} \right\}$$

$$C_3 = \dfrac{(1 + v)}{(1 - v)} \dfrac{A_n v}{2D \lambda_n^4} \times \left\{ (1 - \cosh \lambda_n a)(\sinh \lambda_n a - \lambda_n a) \right.$$

$$+ \dfrac{1}{v} (\sinh \lambda_n a - \lambda_n a \cosh \lambda_n a) \middle/ \dfrac{3 + v}{2} \sinh^2 \lambda_n a$$

$$\left. + \dfrac{1 - v}{2} (\lambda_n a)^2 + \dfrac{2}{1 - v} \right\}$$

$$C_4 = \dfrac{A_n v}{2D \lambda_n^3} \times \left\{ 1 - (1 + v) \sinh^2 \lambda_n a \right.$$

$$- \dfrac{2(1 - v)}{v} \left(\lambda_n a \sinh \lambda_n a + \dfrac{2 \cosh \lambda_n a}{(1 - v)} \right) + (1 - v)(\lambda_n a)^2 \middle/$$

$$\left. (3 + v) \sinh^2 \lambda_n a + (1 - v)(\lambda_n a)^2 + \dfrac{4}{(1 - v)} \right\}$$

4.6. $w_{\max} = w(a/2, b/2) = \dfrac{p_0 a^4}{4 \pi^4 D}$

$$\sigma_{x_{\max}} = \sigma_{y_{\max}} = \sigma(a/2, b/2, \pm h/2) = \pm \dfrac{3 p_0 a^2 (1 + v)}{2 \pi^2 h^2}$$

$$\sigma_{xz} \left(\dfrac{0}{a}, \dfrac{a}{2}, 0 \right) = \sigma_{yz} = \left(\dfrac{a}{2}, \dfrac{0}{a}, 0 \right) = \pm \dfrac{3 p_0 a}{4 \pi h}$$

4.7. $c_1 = 4.07 \times 10^{-3}; \; c_2 = 0.0485$

4.8. $b = 4.58$ ft.; $w_{max} = 0.1485$ in.

4.9. $h = 0.114$ in.; $w_{max} = 0.077$ in.

4.10. $B_{mn} = \dfrac{16p_0}{mn\pi^2}\bigg]_{m,\,n \text{ odd only}} - \dfrac{8p_1(-1)^n}{mn\pi^2}\bigg]_{m \text{ odd only}}$

4.12.   Clamped: $h = 0.332$ in.; Simply supported: $h = 0.358$ in.

4.13.   (a) $h = 0.97$ in.; (b) $h = 0.686$ in.; (c) steel; (d) $h = 0.874$ in.

4.14.   (a) $D\nabla^4 w = -p_0 - kw$; (b) $M_x = V_x = 0$; (c) $M_y = V_y = 0$

4.15.   (a) $C_2 = 0.0833$; $h = 0.430$ in.; $W = 473.1$ lbs.
         (b) $C_2 = 0.119$; $h = 0.514$ in.; $W = 565.1$ lbs.

4.16.   (a) Simply supported: $h = 0.14$ in.; $W_p = 24.41$ lbs.; $W_{TOT} = 44.4$ lbs.
         Clamped: $h = 0.11$ in.: $W_p = 19.8$ lbs.; $W_{TOT} = 59.8$ lbs.
         (b) $h = 0.16$ in.; $W_p = 18.31$ lbs.; $W_{TOT} = 38.31$ lbs.
         (c) $h = 0.09$ in.; $W_p = 47.89$ lbs,; $W_{TOT} = 67.89$ lbs.

4.18.   (a) $h = 0.29$ in.; (b) $h = 0.26$ in.

5.1. $\Delta T = 40 + 300\,z$ (°F); $N^* = 800$ lbs./in.; $M^* = 20$ in. lbs./in.

5.2. $\Delta T = 10 + 8000\,z^2$ (°F); $N^* = 733.3$ lbs./in.; $M^* = 0$ by symmetry

5.5. $\Delta T = 60 + 175\,z + 125\,z^2$ (°F); $N^* = 2467$ lbs./in.;
        $M^* = 93.2$ in. lbs./in.

5.6. $\Delta T = 50 + 60\,z + 30\,z^2$ (°F); $N^* = 120\,E\alpha$ lbs./in.;
        $M^* = 40\,E\alpha$ in. lbs./in.

5.8. $\Delta T = 10 + 220\,z^2$ (°F); $N^* = 16670$ lbs./in.; $M^* = 0$

6.1. $w(0) = -\dfrac{p_1 a^4}{2D}\left[\dfrac{1}{32} - \dfrac{1}{75}\right] = -0.0089\,\dfrac{p_1 a^4}{D}$

6.2. $M_r = M_\theta = M$

6.4. $w(0) = \dfrac{p_0 a^4}{64D}\dfrac{(5 + v)}{(1 + v)}$; $\delta = \dfrac{L_2}{L_1}\,w(0)$

6.5. $w_{max} = w(0) = -\dfrac{p_0 a^4}{64D}$; $\sigma_{r_{max}} = \sigma_r(a,\,\pm h/2) = \pm\dfrac{3p_0 a^2}{4h^2}$

         $\sigma_{\theta_{max}} = \sigma_\theta(0,\,\pm h/2) = \mp\dfrac{3p_0 a^2}{8h^2}\,(1 + v)$

6.6.   Same as 6.5 except $p_0 = -\rho h$.

6.7. $Q_{r_{max}} = \dfrac{p_0(1 - s^2)a}{2s} = \dfrac{p_0(a^2 - b^2)}{2b}$

6.8. $h = 0.5$ in.

6.9.   (a) $h = 0.294$ in.; (b) $w(0) = 0.1095$ in.

6.13. $A + \ln s + Cs^2 + Es^2 \ln s + \dfrac{p_0 a^3 s^4}{64D} = 0$

$$\frac{B}{s} + 2Cs + E[2s \ln s + s] + \frac{p_0 a^3 s^3}{16D} = 0$$

$$-\frac{D}{a}[-B(1 + v) + 2C(1 + v) + (3 + v)E] = 0$$

$$-\frac{4DE}{a^2} - \frac{p_0 a}{2} = 0$$

6.14.   (a) $h = 0.16$ in.; (b) $h = 0.16$ in.

6.15.   (a) $D\nabla^4 w = -p_0 - kw$; (b) $Q_r(a) = M_r(a) = 0$

7.1.   (a) $h = 0.323$ in.; (b) $h = 0.224$ in.

7.2. $N_{x_{cr}} = -\dfrac{2\pi^2 D}{a^2} = -\dfrac{2\pi^2 D}{b^2}$

7.4.   20.23 %

9.1.   (a) $h = 0.506$ in.; (b) $h = 1.21$ in.

9.2.   (a) $P_{cr} = -12EI/L^2$, yes, no.
(b) Yes, yes.

9.4. $N_{x_{cr}} = -D\left[\dfrac{4\pi^2}{a^2} + \dfrac{6(1 - v)}{b^2}\right]$

     Note: The first term is analogous to the result for a clamped-clamped column of unit width; the second term clearly shows the influence of the unloaded edges being simply supported and free.

9.5.   Zero.

9.6.   If $w(x, y) = A\left[1 - \cos\left(\dfrac{2\pi x}{a}\right)\right]\sin\left(\dfrac{\pi y}{b}\right)$ is assumed:

$$N_{x_{cr}} = -\frac{4\pi^2 D}{b^2}\left\{\left(\frac{a}{b}\right)^2 + \frac{3}{16}\left(\frac{a}{b}\right)^3 + \frac{1}{2}\left(\frac{a}{b}\right)\right\}$$

9.7.   If $w(x, y) = A\left[1 - \cos\left(\dfrac{2\pi x}{a}\right)\right]\left[1 - \cos\left(\dfrac{2\pi y}{b}\right)\right]$ is assumed, then

$$N_{x_{cr}} = -\frac{4\pi^2 D}{b^2}\left[\frac{2}{3} + \left(\frac{a}{b}\right)^2 + \left(\frac{b}{a}\right)^2\right]$$

9.8.   If $w(x) = A\left[\dfrac{x^3}{a^2} - \dfrac{2x^2}{a} + x\right]$ is assumed, then

$P_{cr} = -30EI/a^2$

9.9. $w(L/2) = \dfrac{2L^4(2A + c)}{\pi^5 EI}$

$\sigma_{x_{max}} = \sigma_x(L/2) = \mp\dfrac{12(2A + c)L^2}{\pi^3 h^2}$

9.10. $w_{max} = \dfrac{2q_0 L^4}{\pi^5 EI}$

$\sigma_{x_{max}} = \dfrac{12q_0 L^3}{\pi^3 bh^2}$

9.13.   (a) $A = 4q_0 a^4/\pi^5 EI$; (b) 0.385%

9.15. $\omega_1(\text{M.P.E. sol'n}) = 22.792(EI/\rho AL^4)^{1/2}$
$\omega_1(\text{exact sol'n}) = 22.37(EI/\rho AL^4)^{1/2}$

10.1. $\sigma_x(L, +h/2) = -52\,000$ psi.; $\sigma_\theta(L, +h/2) = -15\,600$ psi

10.2. $H = -148.5$ lbs./in.

10.3. $\sigma_\theta\left(\dfrac{\pi}{4\varepsilon}, +h/2\right) = 0.875\, p_0 R/h$; $\sigma\left(\dfrac{\pi}{4\varepsilon}, -h/2\right) = 0.577\, p_0 R/h$

10.4. $w_s = u_{0_p}$; $u_{0_s} = w_p$; $\dfrac{dw_s}{dx} = \dfrac{dw_p}{dr}$; $N_{x_s} = -Q_{r_p}$

$Q_{x_s}(L) = -N_{r_p}$ (or $Q_{x_s}(0) = N_{r_p}$); $M_{x_s} = M_{r_p}$

10.5.   (a) and (b) $\sigma_x\left(\begin{matrix}0\\L\end{matrix}, +h/2\right) = 23\,150$ psi

$\sigma_x\left(\begin{matrix}0\\L\end{matrix}, -h/2\right) = -45\,650$ psi

$\sigma_\theta\left(\begin{matrix}0\\L\end{matrix}, +h/2\right) = 6945$ psi

$\sigma_\theta\left(\begin{matrix}0\\L\end{matrix}, -h/2\right) = -13\,695$ psi

(c) $\left.\begin{matrix}\sigma_x = -11\,250 \text{ psi}\\ \sigma_\theta = -22\,500 \text{ psi}\end{matrix}\right)$ at any location not in a bending boundary layer.

10.6.   (a) $M_0 = \dfrac{\rho(1 - \varepsilon L)}{2\varepsilon^3}$; $Q_0 = \dfrac{\rho(\varepsilon L - 1)}{2\varepsilon^2}$, $M_L = Q_L = 0$

(b) $\sigma_{x_{\max}} = \sigma_x(0, \pm h/2) = \mp \dfrac{3\rho(\varepsilon L - 1)}{h^2 \varepsilon^3}$

(c) $w(L) = 0$

10.7. $M_{L_1} = M_{0_2} = \dfrac{p_i}{4\varepsilon^2} \dfrac{(D_1 - D_2)^2}{(D_1 + D_2)^2 - \frac{1}{2}(D_1 - D_2)^2}$

$Q_{L_1} = Q_{0_2} = -\dfrac{p_i}{2\varepsilon} \dfrac{(D_1 - D_2)^2}{(D_1 + D_2)^2 - \frac{1}{2}(D_1 - D_2)^2}$

11.1. $\sigma_{cr} = -5300$ psi

11.2.   6076 ft.; $\sigma_{cr} = -131\,250$ psi

11.4. $h = 0.09''$

11.11 $P_{cr} = 4.15 \times 10^6$ lbs.

11.12. $P_{cr} = 7.068 \times 10^6$ lbs.

11.13. $P_{cr} = 2.32 \times 10^8$ lbs.

11.14. $L_B = 10$ in.

11.15. $M_{cr} = 5.89 \times 10^7$.-lbs.

11.16. $z = 2375$; $P_{cr} = 52.78$ psi

11.17. $T_{cr} = 2.63 \times 10^7$ in.-lbs.

11.18.   (a) $P_{cr} = -495\,000$ lbs.; (b) $P_{cr} = 1\,130\,000$ lbs.; (c) $L_B = 6.32$ in.

11.19. $M_{cr} = 7.71 \times 10^6$ in. lbs.

11.20. $h = 0.2$ in.

11.27.   (a) 5.66 in.; (b) $\sigma_x = 10\,000$ psi, $\sigma_\theta = 20\,000$ psi; (c) $P_{cr} = 538\,000$ lbs.;
          (e) 138 psi.

Index